P9-DVU-943

DATE DUE

Radiometric Calibration: Theory and Methods

Radiometric Calibration:
Theory and Methods

CLAIR L. WYATT

Electrical Engineering Department
and
Electro-Dynamics Laboratories
Utah State University
Logan, Utah

ACADEMIC PRESS   New York   San Francisco   London   1978
A Subsidiary of Harcourt Brace Jovanovich, Publishers

ACADEMIC PRESS, INC.
111 Fifth Avenue, New York, New York 10003

United Kingdom Edition published by
ACADEMIC PRESS, INC. (LONDON) LTD.
24/28 Oval Road, London NW1 7DX

Library of Congress Cataloging in Publication Data

Wyatt, Clair L
    Radiometric calibration.

    Includes bibliographical references and index.
    1.   Electrooptical devices— —Calibration.
2.   Radiation— —Measurement— —Mathematical models.
I.   Title.
QC673.W9   530'.7   78—4796
ISBN 0—12—766150—6

Contents

Chapter IV   Blackbody Radiation

Chapter V   Geometrical Flux Transfer

Chapter VI   Engineering Calibration

Chapter VII   Standards and Calibration Uncertainty

Chapter XIII **Spectral Calibration**

Chapter XIV **Temporal Response**

Chapter XV **Polarization Response**

Chapter XVI **Practical Calibration of Cryogenic LWIR Systems**

Chapter XVII **Calibration of a Radiometer—A Detailed Example**

Chapter XVIII **Calibration of an Interferometer– Spectrometer—A Detailed Example**

Preface

This book contains an engineering development of the theories and methods of radiometric calibration. It has been written for the engineer and applied physicist concerned with sensor calibration and the interpretation of sensor data.

The scope of this book is necessarily limited to the area of the author's experience, but that includes the radiometric calibration of electrooptical sensors based upon the geometrical transfer of noncoherent radiation.

The first five chapters present an introduction to nomenclature, radiation geometry, and blackbody radiation which serves to simplify the presentation of the calibration theory. Chapters VI through XVIII provide the theory of sensor calibration, giving numerous examples in which laboratory equipment and specific techniques are described. In addition, algorithms are presented for digital computer processing as appropriate for each functional aspect of sensor characterization.

The theory of radiometric calibration, which requires a knowledge of mathematics through integral calculus, is based upon a pragmatic approach. The objective of a measurement is stated in terms of a mathematical model. Based upon this model, the theory is developed, which incorporates the parameters of the practical sensor and which therefore provides considerable insight into the interpretation of the data and into the type of errors that occur.

The subject of radiometry and radiometric calibration is seldom of interest in and of itself, but is generally considered in connection with some other subject. Those who publish the results of their studies tend to group themselves according to their major field of interest, such as astrophysics, aeronomy, meteorology, photometry, illuminating engineering, or optical pyrometry. Each group has developed its own concepts, symbols, and terminology, and publishes in its own journals, thus tending to discourage interdisciplinary communications. These factors have resulted in a fragmentary and limited general development of the theory and methods of calibration.

This book is written to provide a reference work on the subject of electrooptical sensor calibration and to bridge the interdisciplinary communications gap. This is accomplished primarily by a generous use of the units by which the basic radiometric entities can be identified, regardless of the terms and symbols used—the units are given in square brackets following each formulation.

It is not possible to mention all those who have contributed to the contents of this book. However, the author is indebted to Dr. Doran J. Baker, of the Utah State University (USU) faculty and Director of the Electro-Dynamics Laboratories (EDL), to the students who took the electrooptics classes, and to the members of the EDL staff. Also, A. T. Stair and Thomas P. Condron of the Air Force Geophysics Laboratory and Fred E. Nicodemus of the National Bureau of Standards, Washington, D.C., have made substantial contributions, although the author must assume full responsibility for the contents.

Radiometric Calibration: Theory and Methods

CHAPTER

I

Introduction and Objectives

1-1 INTRODUCTION

The electrooptical sensor is a powerful tool with applications that vary from the characterization of the electronic states of atoms to the characterization of distant astronomical objects—from the smallest to the largest objects known to man. It has been used in military, industrial, and scientific applications too numerous to outline here [1, 2], and they will continue to play an important role in the modern world.

The radiometer (or photometer) and the spectrometer (including both sequential and multiplex types) are given prime emphasis here, but the general theory and practice apply also to spatial scanners, imaging systems, and polarimeters which are given less emphasis.

Radiometry (or spectroscopy) is concerned with the transfer of optical radiation between a target source and its associated background, through the intervening medium, to a receiver or detector of optical radiant energy. The problem is to determine the quantity and quality of energy or flux flowing in a beam of radiation. This is illustrated in Fig. 1-1 where the target source is shown imbedded in a background and an intervening atmosphere and illuminated by a source. The target, the background, and the atmosphere are represented as sources of reflected and emitted radiant energy. Some of the rays reflected and scattered from these sources reach the collecting aperture of the sensor.

1

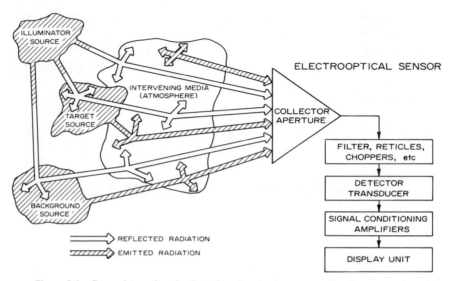

Figure 1-1   General transfer of reflected and emitted energy or flux from a target and its background through an intervening media to an optical sensor which provides an output proportional to the quantity and the quality of the input radiation. (Adapted from F. E. Nicodemus, Radiometry. *In* "Optical Instruments," Part 1 (R. Kingslake, ed.), Applied Optics and Optical Engineering, Vol. 4, p. 288. Academic Press, New York, 1967.)

The measurement problem can be considered as the problem of the characterization of the target using an electrooptical sensor as a remote detector of radiant flux. Characterization of a target means the determination of the target attributes such as the size, shape, or location of the target, its temperature, radiant properties, reaction rate, energy levels, etc.

Most of the target attributes cannot be measured directly [3] by remote sensing of radiant flux, but must be inferred from the instrument response to the flux incident upon the sensor aperture [3, p. VII-9]. For the limited case of noncoherent and noninterfering radiation, the target can be characterized in terms of four nearly independent domains: spatial (geometrical extent), temporal (time variations), spectral (distribution of energy of flux as a function of wavelength or optical frequency), and polarization.

It follows, therefore, that the calibration must also characterize the sensor response in these four domains. Large measurement errors can result from incomplete characterization of the out-of-band and/or the off-axis response of a sensor. For example, an antiaircraft missile guidance system may track the sun rather than the enemy aircraft; a rocket-borne radiometer may respond to thermal emissions of the earth rather than the hydroxyl emissions of the atmosphere in the earth limb; or a spectrometer designed to measure the Balmer lines in the ultraviolet may respond to the solar continuum. The degree to which this characterization of the optical sensor must be carried out depends on the attributes of the target and its background.

Target discrimination is a problem of the detection of a signal that is imbedded in noise. Discrimination can be accomplished by fully utilizing the characteristics of the target and the background.

The measurement problem that the sensor is designed to solve can also be considered as the problem of determining the quantity (the absolute or relative intensity) and the quality (within certain geometrical, spectral, and polarization limits) of the radiation scattered and emitted by the target source.

The calibration problem is to determine a functional relationship between the target source flux and the display unit output. Generally this is expressed as a *responsivity* in units of volts per unit flux, although the output might be current, count rate, deflection of a pen, density of an emulsion, etc.

The quality of the measurement is the most difficult aspect of calibration. As depicted in Fig. 1-1, the instrument aperture is bombarded with unwanted flux which arrives from outside the instrument field of view, such as the sun, earth, lights, etc. The sensor output for a spatially pure measurement is a function of the radiant flux originating from the target (within the sensor field of view) and is completely independent of any flux arriving at the instrument aperture from outside this region. Thus, in this book, the calibration of the spatial response, or field of view of a sensor, is considered to be a problem of *spatial purity*.

The instrument aperture is also bombarded with unwanted flux which is out-of-band, that is, outside of the spectral band of interest. The sensor output for a spectrally pure measurement is a function of the radiant flux originating from within the sensor spectral bandpass and is completely independent of any flux arriving at the instrument aperture from any spectral region outside the bandpass. Thus, in this book, the characterization of the spectral response of a sensor is considered as a problem of *spectral purity*.

The flux emanating from the target may be changing with time and may be polarized. The instrument may be moving or, for other reasons, may be time limited, and may be sensitive to the polarization of the incident flux. The time and polarization characteristics of the target may be considered as target attributes for discrimination against the background, may be related to physical processes in materials, or may prove to limit the detection of the target or of its characterization in the spatial or spectral domain.

Optically pure measurements are only approximated with practical instruments which are always nonideal. Consequently, the interpretation of field data is subject to some uncertainty and is often dependent on assumptions that must be made about the target.

The theory of radiometric calibration, as applied to sensors used to characterize remote target sources, is not well understood by many engineers and physicists engaged in measurement programs. However, the correct interpretation of field data is dependent on an understanding of the geometry of radiation and the theory and practical limitations of calibration. The "ideal" approach to calibration is often not possible because of the limitations of existing and/or available equipment. Consequently, many compromises in the "ideal" method must often be made. The nature of these compromises depends on the type of sensor and the resources (facilities, manpower, monies, etc.) that are available.

1-2   CALIBRATION OBJECTIVES

The general objective of the calibration of electrooptical instrumentation, for the measurement of remote radiant sources, is to obtain a functional relationship between the incident flux and the instrument output. The functional relationship is generally expressed as a mathematical equation (or a tabulation of values) which gives the magnitude of the radiant entity of interest Φ, as a function of the instrument output Γ as

$$\Phi = f(\Gamma). \tag{1-1}$$

All systems are not necessarily linear by design or by nature. Some detectors–transducers are inherently nonlinear. Also some nonlinear schemes are occasionally used to extend the dynamic range of systems when the costs of telemetry are too great to make use of multiple linear channels. However, some aspects of the sensor calibration must be evaluated with linearized data.

Equation (1-1) takes the form of the product of the instrument output Γ with a constant known as the *inverse responsivity* $1/R$, namely,

$$\Phi = \Gamma(1/R) \tag{1-2}$$

which is measured in units of flux per unit output for linear systems in which the *offset error* (the output for zero flux input) is zero. Inherent in constant R is the complete characterization of the instrument field of view, spectral bandpass, time constant, and polarization, as well as the gain of any associated electronic amplifiers, recorders, emulsions, counters, etc., that are used as signal-conditioning output systems. Equation (1-2) is the *calibration equation* that is used to convert an instrument output to the radiant entity of interest.

A major objective of the calibration of electrooptical sensors is as follows [4]:

The calibration of an instrument for a specific measurement should be provided in such a way as to make the measurement independent of the instrument.

This means that when a particular physical entity is to be measured at different times, places, or with different instruments, the results should always be the same. It also means that attributes inferred by the measurements are target attributes and not instrument attributes.

Unfortunately, the complexity of the physical conditions of the instrument and source parameters existing at the time of the measurement make this ideal of calibration somewhat difficult, if not impossible, to obtain. A basic rule of good performance that tends toward this ideal of calibration is as follows [4, p. 19]:

The calibration should be conducted under conditions which reproduce, as completely as possible, the conditions under which the measurements are to be made.

This rule can be observed only in a limited way because it implies a foreknowledge of the source characteristics which would render the measurement unnecessary. It is usually possible to do quite well as far as the spatial parameters are concerned. For example, it is possible to calibrate with an extended source for extended source measurements or conversely with a point source for point source measurements. However, it is much more difficult as far as the spectral parameters are concerned. This is because the instrument-relative power spectral response weights the incident source flux in a way difficult to qualify. The only method by which the problem can be avoided is to use a calibration source that has exactly the same power spectral density as the source being measured [4, p. 19]. Calibration sources are generally limited to the spectral characteristics of a blackbody, so it is often impossible to match the spectra.

Any report of measurements of a physical entity should include a statement of errors [4, p. 2]. The errors may be qualified in terms of the *precision* and the *accuracy* of the measurement [5] as detailed in a later chapter. The statement of errors should include the calibration procedure followed, identification of absolute source standards, and measured reproducibility. The instrument parameters, on which the measurement depends, should also be included in the report.

The calibration of the instrument requires a functional set of data concerning the spectral, spatial, temporal, and polarization characteristics of the instrument. Before obtaining calibration information relative to these major parameters, it is necessary to investigate the linearity and the background noise level in the instrument itself. The functional parameters whose collective ascertainment is the objective of electrooptical sensor calibration are categorized in the accompanying tabulation. Each of the calibration parameters shall be defined and appropriate practical techniques by which each is determined shall be discussed in this book. The order or presentation corresponds roughly to that logically undertaken in the calibration of a sensor.

Parameters	Parameters
Dark-noise analysis	Spectral calibration
standard deviation	scan position
mean (offset error)	spectral resolution
Linearity analysis	relative spectral responsivity
Field-of-view calibration	spectral purity
near field	Absolute calibration
far field	Temporal calibration
	Polarization response

Chapters 17 and 18 give detailed step-by-step illustration of the calibration of a radiometer and an interferometer–spectrometer. These are intended to provide a "this-is-how-you-do-it" guide.

Although the literature contains detailed theoretical developments on radiation transfer, as referrred to throughout this book, there exists little practical information on calibration. An exception is the recent work by Condron [6].

REFERENCES

1 Special Issue on Infrared Physics and Technology. *Proc. IRE*, **47**, 1413–1700 (1959).
2 M. R. Holter, S. Nudelman, G. H. Suits, W. L. Wolfe, and G. J. Zissis, "Fundamentals of Infrared Technology." Macmillan, New York, 1962.
3 "University of Michigan Notes for a Program of Study in Remote Sensing of Earth Resources," February 14, 1968–May 3, 1968, Contract NAS 9-7676, p. I-1. NASA, MSC, Houston, Texas (1968).
4 F. E. Nicodemus and G. J. Zissis, "Report of BAMIRAC—Methods of Radiometric Calibration," ARPA Contract No. SD-91, Rep. No. 4613-20-R (DDC No. AD-289, 375), p. 5. Univ. of Michigan, Infrared Lab., Ann Arbor, Michigan (1962).
5 H. J. Kostkowski and R. D. Lee, Theory and Methods of Optical Pyrometry. *Nat. Bur. Stand. (U.S.), Monogr.* No. 41 (1962).
6 T. P. Condron, Calibration techniques. *In* "Spectrometric Techniques" (G. A. Vanasse, ed.), Vol. 1, Ch. 7. Academic Press, New York, 1977.

II

Definitions

2-1 THE ELECTROMAGNETIC SPECTRUM

Light was, according to Maxwell, a component of the *electromagnetic spectrum* (Fig. 2-1). All the waves shown in the figure are electromagnetic in nature, and have the same speed c in free space. They differ in wavelength (frequency and wave number) only. However, the sources that give rise to these radiations and the various types of instruments used to measure them are very different.

The electromagnetic spectrum has no definite upper or lower limit. The labeled regions of Fig. 2-1 represent wavelength intervals within which a common body of experimental techniques, such as detection techniques, exist. All such regions overlap.

In this book the basic units used are those of the International System (SI) consisting of the meter, kilogram, and second (see Appendix A). However, exception is taken in the case of wavelength and wave number as used to designate the spectrum.

The symbol λ is used to designate *wavelength* which is defined as the distance between two adjacent points in a wave having the same phase. The most common unit for wavelength is the micrometer (μm), where μ is a prefix that represents 10^{-6} (see Appendix B for a list of prefixes for use with the SI system). Alternative units for wavelength are the nanometer (nm), where n represents 10^{-9}, and the angstrom (Å), which is defined as 10^{-10} m.

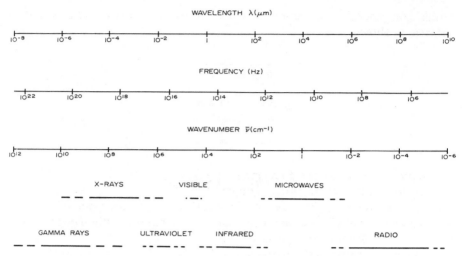

Figure 2-1 The electromagnetic spectrum. (Note: The scale is logarithmic.)

The symbol used to represent wave number is $\bar{\nu}$, and might be defined as the number of wavelengths that would, at an instant, occupy a space 1 cm in length. The unit for wave number must therefore be reciprocal centimeters (cm^{-1}) where

$$\bar{\nu} = \nu/c = 1/\lambda \quad [cm^{-1}] \qquad (2-1)$$

with ν the frequency in hertz (cycles per second).

The spectral ranges considered in this book are categorized on the basis of distinct detection techniques as shown in the accompanying tabulation.

Ultraviolet (uv)	1800–4000 Å
Visible (vis)	4000–7000 Å
Near infrared (NIR)	7000–10,000 Å
Short-wave infrared (SWIR)	1–5 μm
Long-wave infrared (LWIR)	5–25 μm

2-2 INSTRUMENTATION TYPES

There are a number of terms used to describe radiometric devices in the literature [1]. These terms are not used consistently but do convey some indication of the type of measurement intended for that instrument. These terms are generally composed of combinations of certain prefixes and suffixes, as shown:

Prefix	Suffix
radio	meter
photo	graph
spectro	scope

radio   refers to electromagnetic radiation in general,

photo   refers to visible radiation,

spectro refers to the division of the radiation into components,

meter   implies a measurement is indicated, but the method of presentation is not specified,

graph   implies that a measurement is recorded in some graphic form,

scope   implies that the radiation is to be viewed by the eye through the device.

Thus *radiometer* is a general term that represents any device used to measure radiation; however, the term generally implies the absolute measurement of radiant flux over a specific wavelength interval. Additional prefixes may be used to specify the spectral region that the particular instrument measures, such as infrared radiometer, ultraviolet radiometer, and microwave radiometer. A *photometer* is an instrument for use in the visible range, and it is implied that the response is adjusted to give measurements in terms of the visual effect.

In a similar manner, a *spectrometer* is a general term that represents any device used to measure the radiation at a selected wavelength or wavelengths. A spectrometer measures the distribution of radiant flux as a continuous function of wavelength (frequency or wave number) throughout a limited region of the electromagnetic spectrum which is defined here as the *free spectral range*. This distribution is known as the *power spectral density function* or simply the *spectrum*. Figure 2-2 shows the airglow spectrum of the hydroxyl molecule obtained with a field-widened Michelson interferometer–spectrometer [2].

A *spectrograph* is a spectrometer that produces a graphic recorded spectrum as an inherent part of its function. Often, the term "spectrograph" is used to imply that the recording is done on photographic film placed in the image plane. Such a recording is called a *spectrogram*. A *recording spectrometer* is a spectrometer that produces a graphical plot of power as a function of wavelength rather than as a photograph of the spectrum.

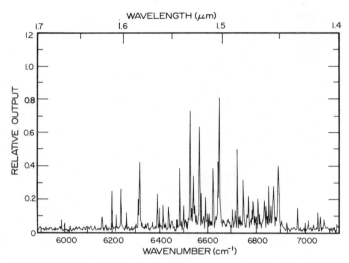

Figure 2-2   The airglow spectrum of the hydroxyl molecule obtained with the Utah State University field-widened interferometer–spectrometer.

There are several other types of spectrometers. A *spectroscope* is a spectrometer that provides for direct observation with the eye. A *spectroradiometer* is a spectrometer that has been calibrated for absolute measurements of the spectrum. Such an instrument functions in a way that is roughly equivalent to *n* radiometers, where *n* is the number of resolution elements within the free spectral range of the spectrometer. An absolute calibration of a spectroradiometer is very difficult to obtain. Often a spectrometer is used only to measure the *relative* power spectral density; such data are used to identify emission or absorption features and the wavelength at which they occur.

A simple spectrometer sequentially scans through the free spectral range consisting of *n* wavelength increments. More sophisticated "multiplex" spectrometers, such as a *Michelson interferometer–spectrometer* [2] or the *Hadamard multiplex spectrometer* [3] scan so that all wavelengths within the free spectral range are sampled at the same time and encoded so they can be separated by an inverse transform procedure.

Radiant energy can be analyzed into its component wavelengths by dispersion techniques or by interferometric techniques. A dispersion spectrometer is one that employs either a prism or a grating to separate the energy into its components. An interferometric spectrometer (Fourier spectroscopy) is based on the Michelson interferometer.

In this book the terms *electrooptical sensor, radiometer,* and *spectrometer* are often used interchangeably. This is because most concepts dealing with

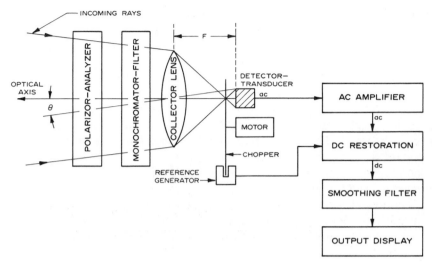

Figure 2-3 Basic absolute radiometer–spectroradiometer.

radiation transfer and calibration technique apply equally well. The radiometer can be considered mathematically as a special case of the spectroradiometer that is temporarily fixed at one wavelength.

The basic design of an absolute radiometer–spectroradiometer is given in Fig. 2-3. The rays of radiant flux, emanating from a remote target, pass through the polarizer–analyzer, the monochromator–filter, and the collector lens, and are imaged upon the detector–transducer. The radiant flux is coded by the chopper in such a way that the output of the detector is an alternating electrical signal that is amplified, dc restored, filtered, and presented for display. The characterization of the remote source in the domains of polarization, spectral, spatial, and temporal is provided in the following way.

The polarization analyzer provides for the selection of the mode of polarization of the sensor. The analyzer can be rotated to provide for a continuous presentation of the incoming radiant flux (in time). The monochromator–filter determines the spectral bandpass of the sensor by transmitting only those rays desired and by blocking the rays that are outside the bandpass. The sensor functions as a radiometer if the monochromator is fixed at one wavelength. However, the system functions as a spectroradiometer if the monochromator scans continuously (in time) through a free spectral range of wavelengths (or wave numbers). The lens forms the collecting optics with a circular aperture of a given area. The detector–transducer forms a circular field stop located at the lens focal length F. The radius of the field stop determines Θ the half-angle edge of the field of view. The sensor can obtain a line scan (or even a raster) provided that the optical axis is scanned geometrically by physically moving the entire sensor or through the use of tilting or rotating mirrors. Thus the field stop and the lens determine the instantaneous spatial response or field of view of the sensor.

Absolute levels of flux are defined with respect to absolute zero; thus the level of flux incident on the detector must be compared with a reference of known level. The chopper, pictured in Fig. 2-3, provides a zero reference by periodically blocking the incoming flux, provided that its own radiation is negligibly small. There are other methods, including the use of internal standard sources, to obtain absolute measurements. The reference signal generator provides for dc restoration of the electrical ac signal that results from chopping the incoming flux.

The low-pass filter determines the instrument temporal response (time constant) and limits the electrical noise in the output.

Finally, the output display provides the sensor output in some convenient form for analysis.

2-3 THE RESOLVING POWER OF A SPECTROMETER

The resolving power of a spectrometer is given by

$$\mathscr{R} = \lambda/\Delta\lambda = \bar{v}/\Delta\bar{v}, \tag{2-2}$$

where $\Delta\lambda$ or $\Delta\bar{v}$ is the width of the incremental interval. To convert from wavelength to wave number or the converse it is necessary to differentiate Eq. (2-1). When the resolution is high, $d\lambda$ and $d\bar{v}$ can be replaced by $\Delta\lambda$ and $\Delta\bar{v}$ giving

$$\Delta\bar{v} = -(1/\lambda^2)\,\Delta\lambda \quad [\text{cm}^{-1}]. \tag{2-3}$$

where the wavelength λ must be entered in centimeters (cm).

2-4 SPECTROMETER DATA PRESENTATION

There is a question as to what is the optimum functional presentation of optical spectra [4]. The concern is with the choice of the independent variable used to designate the different "colors" of the spectrum. The choice is among wavelength, wave number, frequency, and energy.

The criterion used here as optimum is based on the width of the resolution element: *It is desirable to maintain the magnitude of the resolution element as constant throughout the plot of the entire spectrum.* As shall be shown, a prism or grating spectrometer has a constant resolution when the spectrum is plotted as a function of *wavelength*, and an interferometer–spectrometer has a constant resolution when the spectrum is plotted as a function of *wave number*. Furthermore, it is shown that for a circular-variable filter spectrometer (CVF) or wedge filter, based on the criteria of a constant resolution element, there does not exist an optimum way to plot the spectrum.

In a *prism spectrometer* each color in the incident beam is deflected through a definite angle θ, determined by the wavelength of that color and the index of refraction of the prism. The dispersion [5] of a prism spectrograph is

$$dx/d\lambda = F(b/a)\,(dn/d\lambda) \quad [\text{m } \mu\text{m}^{-1}], \tag{2-4}$$

where x is the output, F the system focal length, b/a a constant of the prism geometry, n the index of refraction, and λ the wavelength.

Equation (2-4) shows that the dispersion tends to be a constant function of the wavelength of the radiant power (neglecting changes in the index of refraction n); therefore, the width of the resolution element also tends to be a constant function of wavelength and it is appropriate to plot power spectral density as a function of wavelength.

The dispersion of a *grating spectrometer* [6] is given by

$$dx/d\lambda = mF/(d \cos \theta) \quad [\text{m } \mu\text{m}^{-1}], \tag{2-5}$$

where m is the order number ($m = 0, 1, 2, \ldots$), d the distance between grating rulings, θ the diffraction angle, and F the focal length. Here, as with the prism spectrometer, it is appropriate to plot power spectral density as a function of wavelength since the width of the resolution element is a constant function of wavelength.

In a *Michelson interferometer*, each monochromatic element in the incoming radiant spectrum is converted or transformed into an electrical frequency f according to the equation [7]

$$f = v\bar{v} \quad [\text{Hz}], \tag{2-6}$$

where v is the rate of change of path difference (centimeters per second) and $\bar{v}$ the wave number (reciprocal centimeters). The wave number is given by

$$\bar{v} \hat{=} v/c = 1/\lambda \quad [\text{cm}^{-1}], \tag{2-7}$$

where v is the optical frequency and c the speed of light. Thus the output frequency f is also proportional to the optical frequency.

It is helpful to differentiate Eq. (2-6), treating $\bar{v}$ as an input variable and f as the output variable in order to calculate what is roughly equivalent to dispersion for the interferometer, which gives

$$df/d\bar{v} = v \quad [\text{Hz cm}]. \tag{2-8}$$

Thus it is appropriate to plot power spectral density as a function of wave number since, in this case, the width of the element is a constant function of wave number.

For a circular-variable (wedge) filter (CVF) the center wavelength of the optical bandpass is a linear function of the polar angle on the filter. The width of the resolution element is also proportional to the center wavelength and, hence, to the scan position. Consequently, the width of the resolution element is linearly related to wavelength and to wave number, and is therefore not a constant of either. Thus there is no optimum way to plot the spectrum obtained from a CVF spectrometer [8].

REFERENCES

1   Anonymous, Infrared measuring instruments. *In* "Handbook of Military Infrared Technology" (W. L. Wolfe, ed.), Ch. 19. Off. Nav. Res., Washington, D.C., 1965.

2   R. J. Bell, "Introductory Fourier Transform Spectroscopy." Academic Press, New York, 1972.

3   R. N. Ibbett, D. Aspinall, and J. F. Grainger, Real-time multiplexing of dispersed spectra in any wavelength region. *Appl. Opt.* **7**, 1089 (1968); see also E. D. Nelson and M. L. Fredman, Hadamard spectroscopy. *J. Opt. Soc. Am.* **60**, 1664 (1970).

4   D. J. Baker and W. L. Brown, Presentation of spectra. *Appl. Opt.* **5**, 1331 (1966).

5   M. Born and E. Wolf, "Principles of Optics," p. 180. Macmillan, New York, 1964.

6   D. Halliday and R. Resnick, "Physics," p. 1040. Wiley, New York, 1963.

7   E. V. Loewenstein, Fourier spectroscopy: An introduction. *Aspen Int. Conf. Fourier Spectrosc.* Spec. Rep. No. 114, AFCRL-71-0019, p. 4. Air Force Cambridge Res. Lab., L. G. Hanscom Field, Bedford, Massachusetts (1970).

8   C. L. Wyatt, Infrared spectrometer: Liquid-helium-cooled rocketborne circular-variable filter. *Appl. Opt.* **14**, 3086 (1975).

CHAPTER

III

Radiometric Nomenclature

3-1 INTRODUCTION

Radiometry and the theory of radiometric calibration are seldom of interest in and of themselves, but are generally considered in connection with some other subject. Those who publish the results of their studies tend to group themselves according to their major field of interest such as astrophysics, aeronomy, meteorology, photometry, biological chemistry, illuminating engineering, or optical pyrometry. Many research centers have developed adequate calibration techniques, yet the lack of interdisciplinary communication has resulted in numerous nearly independent and confusing systems of nomenclature for terms, symbols, and units [1]. Thus each group has developed its own concepts, symbols, and terminology and publishes in its own journals, thus further tending to discourage interdisciplinary communication.

The standards adopted by the International Commission on Illumination [2] CIE, which have also been adopted by the National Bureau of Standards [3], *Applied Optics*, the *Journal of the Optical Society of America* [4], the *Proceedings of the IRIS* (*Infrared Information Symposium*), the Illuminating Engineering Society [5], and the American Society of Mechanical Engineers [6] are given in Table 3-1. The CIE standard is used in this book because of its acceptance by numerous standardizing agencies and because of its wide appeal. However, the CIE standard is incomplete;

for example, the term *radiance* has the units of watts per square meter steradian ($W\ m^{-2}\ sr^{-1}$), but there exists no term corresponding to the units of quanta per second square meter steradian ($q\ sec^{-1}\ m^{-2}\ sr^{-1}$). More importantly, there is no term for the general entity which has the units of flux per square meter steradian ($\Phi\ m^{-2}\ sr^{-1}$), where Φ is a general term for any flux that obeys the laws of the geometry of radiation such as watts, quanta per second, lumens, rayleighs, etc. Furthermore, there are numerous research groups who deal with the basic concepts of radiation transfer who do not conform to the CIE standard.

TABLE 3-1

Nomenclature[a]

Terms[b]	Symbols	Units	Terms[b]	Symbols	Units
Flux	Φ	Φ	Luminous exitance	M_v	$lm\ m^{-2}$
Radiant flux	Φ_e	W	[—]	M_p	$q\ sec^{-1}\ m^{-2}$
Luminous flux	Φ_v	lm			
[—]	Φ_p	$q\ sec^{-1}$	[—]	E	$\Phi\ m^{-2}$
			Irradiance	E_e	$W\ m^{-2}$
[—]	L	$\Phi\ m^{-2}\ sr^{-1}$	Illuminance	E_v	$lm\ m^{-2}$
Radiance	L_e	$W\ m^{-2}\ sr^{-1}$	[—]	E_p	$q\ sec^{-1}\ m^{-2}$
Luminance	L_v	$lm\ m^{-2}\ sr^{-1}$			
[—]	L_p	$q\ sec^{-1}\ m^{-2}\ sr^{-1}$	[Intensity]	I	$\Phi\ sr^{-1}$
			Radiant intensity	I_e	$W\ sr^{-1}$
[Exitance]	M	$\Phi\ m^{-2}$	Luminous intensity	I_v	$lm\ sr^{-1}$
Radiant exitance	M_e	$W\ m^{-2}$	[—]	I_p	$q\ sec^{-1}\ sr^{-1}$

[a] Adapted from the American National Standard Nomenclature for Illuminating Engineering. These terms are essentially the same as those of ANSI Z7.1-1967 and those adopted by the CIE internationally.

[b] Dashes represent missing terms in the CIE nomenclature.

It is noted that at the time of this writing that there has been a continuous search for acceptable *terms* to describe the entities associated with geometrical radiation of flux. Jones' *phluometry* scheme has received considerable attention [7, 8]. Phluometry, defined as the geometry of radiation, incorporates terms which convey the geometrical properties of the entities envolved. The principle terms suggested in recent communication [7] are almost self-explanatory:

sterance	related to the solid angle (steradian)
areance	related to an area
pointance	related to a point

The use of the terms "sterance," "areance," and "pointance" provide the degree of generalization desired for a book on calibration techniques. Furthermore, the use of these geometrical terms should bridge the communication gap for those who are presently using nomenclature schemes other than that of the CIE standard and at the same time provide a logical extension of terms for those entities not included in the CIE standard.

The new terms have been suggested as alternate standard terms to be used in conjunction with the CIE terms. In this book the terms are combined as:

sterance	[radiance, luminance]
areance	[irradiance, illuminance]
pointance	[intensity]

and so on. Thus the new terms are used in this book as those most appropriate with which to illustrate the geometry of radiation, however, the CIE terms (when defined for a particular entity) will be given immediately following in square brackets. Also, the entity can be identified by the units that are generally given throughout this book. Thus it is hoped that the book can be easily read by anyone interested in electrooptical sensor calibration regardless of their background.

The International System of Units (SI) will be used for dimensional analysis. However, there are some exceptions which are noted. Appendix A gives a list of the basic SI units.

3-2   ENTITIES BASED ON FLUX, AREA, AND SOLID ANGLE

The units given in Table 3-2, which lists the radiometric entities of interest, are based on the elementary concepts of flux, area, and solid angle.

Flux is defined as any quantity that is propagated or spatially distributed according to the laws of geometry or the geometry of radiation (phluometry). Examples are radiant energy, radiant power, visible light or luminous flux, electromagnetic quanta or photons, spectral radiant power, entropy, etc. The subscripts e, v, p, used in Table 3-2 refer to energy rate (watts), visible (lumen), and photon (quanta) flux, respectively, and are often not used when it is clear from the context in which they appear [9].

In phluometry the most complete and basic function is the geometrical distribution of flux with respect to position and direction—the flux per unit projected area and solid angle (flux per square meter steradian), or flux per unit throughput (Φ/T). The CIE standard retained the terms "radiance"

TABLE 3-2

Basic Radiometric and Photometric Entities

Terms[a]	Symbols	Units
Flux (watts, lumens, quanta, etc.)	Φ	Φ
Radiant flux (watts)	Φ_e	W
Luminous flux (lumens)	Φ_v	lm
Photon flux (quanta per second)	Φ_p	$q \ sec^{-1}$
Sterance [—] (positional–directional)	L	$\Phi \ m^{-2} \ sr^{-1}$
Radiant sterance [radiance]	L_e	$W \ m^{-2} \ sr^{-1}$
Luminous sterance [luminance]	L_v	$lm \ m^{-2} \ sr^{-1}$
Photon sterance[b] [—]	L_p	$q \ sec^{-1} \ m^{-2} \ sr^{-1}$
Areance [exitance] (positional)	M	$\Phi \ m^{-2}$
Radiant areance [radiant exitance]	M_e	$W \ m^{-2}$
Luminous areance [luminous exitance]	M_v	$lm \ m^{-2}$
Photon areance [—]	M_p	$q \ sec^{-1} \ m^{-2}$
Areance [—] (positional)	E	$\Phi \ m^{-2}$
Radiant areance [irradiance]	E_e	$W \ m^{-2}$
Luminous areance [illuminance]	E_v	$lm \ m^{-2}$
Photon areance [—]	E_p	$q \ sec^{-1} \ m^{-2}$
Pointance [intensity] (directional)	I	$\Phi \ sr^{-1}$
Radiant pointance [radiant intensity]	I_e	$W \ sr^{-1}$
Luminous pointance [luminous intensity]	I_v	$lm \ sr^{-1}$
Photon pointance [—]	I_p	$q \ sec^{-1} \ sr^{-1}$

[a] Dashes represent missing terms in the CIE nomenclature.

[b] The *rayleigh* is also a measure of photon sterance (see Section 3-10).

(watts per square meter steradian) and "luminance" (lumen per square meter steradian) and provide no equivalent term for photon flux. Jones suggested the general term *sterance* which by the use of suitable modifiers may be made to apply to any flux. Thus *radiance* becomes *radiant sterance*, *luminance* becomes *luminous sterance*, and the photon quantity for which no term is given becomes *photon sterance*.

The *average sterance* [*radiance, luminance*] L_{av} of a source is the ratio of the total flux to the product of the projected area and the solid angle. The limiting value of the average sterance as both area and solid angle are reduced is the *sterance* L at a point in a direction. The sterance is a measure of the flux of a source per unit area per unit solid angle in a particular direction, and is $L = d^2\Phi/(d\omega \cos \theta \ dA) = d\Phi/dT$ which has the units of flux per square meter steradian.

It is implicit in the definition of sterance that the area is taken as the projected area of the source in the direction that the radiation is directed or being measured. This allows a meaningful measure of the sterance when the source area is not known. If the projected area is used, rather than the real area, the sterance is independent of direction for sources that obey Lambert's cosine law.

The geometrical distribution of flux with respect to position (surface) is the flux per unit area (flux per square meter). The CIE standard includes two entities to differentiate between emitted and incident flux. Emitted flux is given the general term of *exitance* which is modified to provide *radiant exitance* and *luminous exitance*. The terms of *irradiance* and *illumination* are retained in the CIE standard for incident flux. The general term *incidance* was suggested by Jones [8] for incident flux. However, the term *areance* has more recently been suggested [7, pp. G183–G187] as a general term for both emitted and incident flux. Thus *radiant exitance* would be *radiant areance*; also irradiance would be *radiant areance*.

The *average areance* [*exitance*] M_{av} of a source is the ratio of the total radiant flux Φ to the total area of the source. However, the limiting value of the average areance of a small portion of the source as the area is reduced to a point is the *areance M* of the source at a point. The areance is a measure of the flux radiated into a hemisphere per unit area of the source, and is $M = d\Phi/dA$ where A is the area and M has the units of flux per square meter.

The *average areance* [*irradiance, illuminance*] E_{av} is the ratio of the total flux to the total area of the incident surface, and is a measure of the flux per unit area incident on a surface. The limiting value of the average areance, as the area is reduced to a point, is the *areance E* at that point. The areance is a measure of the incident flux per unit area of surface and is $E = d\Phi/dA$, where A is the area and E has the units of flux per square meter.

The geometrical distribution of flux with respect to direction is flux per unit solid angle (flux per steradian). The CIE standard retains the term *intensity* which has been objected to on the grounds that it has so many other meanings [10]. The term *pointance* has been suggested as a general term [7, pp. G183–G187]. Thus *radiant intensity* becomes *radiant pointance*.

The *average pointance* [*intensity*] I_{av} of a source is the ratio of the total flux radiated by a source to the total solid angle about the source. For an isotropic source (radiating equally in all directions), the flux is radiated into 4π sr (into a sphere) and for a flat surface, the flux is radiated into 2π sr (into a hemisphere). The limiting value of the average pointance, as the solid angle is reduced in value about a particular direction, is the *pointance I* in that direction. Pointance is a measure of the flux radiated by a source per unit solid angle in a particular direction, and is $I = d\Phi/d\omega$, where ω is the solid angle and I has the units of flux per steradian.

The CIE terms of exitance, intensity, and radiance or luminance are usually thought of as having reference to a source; irradiance or illumination on the other hand are considered as having reference to a receiver. However, these concepts can also be applied within a radiation field away from sources or receivers. If a barrier containing an aperture is placed in a radiant field, it has the properties of a source for the flux leaving the aperture and a receiver for the flux incident upon it. The radiation field could be defined by reducing the aperture to a point; then there would be a meaningful measure of the sterance, pointance, and areance for that point—or of the *field*. Thus there is no more fundamental reason for distinguishing between incoming or outgoing flux per unit area (flux per square meter), than there is for any other entity. The three general terms *sterance* (flux per square meter steradian), *areance* (flux per square meter) and *pointance* (flux per steradian) embrace basic and unique properties of geometrical radiation.

The utility of considering these entities as field entities [11] will become evident in later sections. These field entities can be measured by placing a properly calibrated radiometer at the point in the field.

3-3   ENTITIES BASED ON FLUX, VOLUME, AND SOLID ANGLE

A set of terms, symbols, and units is required for a clear understanding of propagation through optically thin gases in which emission and/or scattering are taking place. The entities not included in Table 3-2 are listed in Table 3-3. In real situations there can be attenuation (scattering out of the ray) and generation (scattering into the ray), as well as emission along a ray. Jones proposed an entirely new term *sterisent* L^* (flux per cubic meter steradian) to cover these effects.

Sterisent can be viewed as the sterance [radiance, luminance] generated per unit length along the ray (flux per square meter steradian/meter = L/meter) which is a path function, or, alternatively, it can be viewed as the

TABLE 3-3

Additional Radiometric Entities

Terms	Symbols	Units
Sterisent (positional–directional, path function)	L^*	$\Phi\ \mathrm{m}^{-3}\ \mathrm{sr}^{-1}$
Radiant sterisent	$L_e{}^*$	$\mathrm{W}\ \mathrm{m}^{-3}\ \mathrm{sr}^{-1}$
Luminous sterisent	$L_v{}^*$	$\mathrm{lm}\ \mathrm{m}^{-3}\ \mathrm{sr}^{-1}$
Photon sterisent	$L_p{}^*$	$\mathrm{q}\ \mathrm{s}^{-1}\ \mathrm{m}^{-3}\ \mathrm{sr}^{-1}$
Volume emission rate (positional, volume)	σ	$\Phi\ \mathrm{m}^{-3}$
Radiant volume emission rate	σ_e	$\mathrm{W}\ \mathrm{m}^{-3}$
Luminous volume emission rate	σ_v	$\mathrm{lm}\ \mathrm{m}^{-3}$
Photon volume emission rate	σ_p	$\mathrm{q}\ \mathrm{s}^{-1}\ \mathrm{m}^{-3}$

geometrical distribution of flux with respect to volume and direction—the flux per unit volume and solid angle (flux per cubic meter steradian).

The *average sterisent* L_{av}^* of a source is the ratio of the total flux to the product of volume and solid angle. The limiting value of the average sterisent as both volume and solid angle are reduced is the *sterisent* L^* of the source at a point and in a direction. The sterisent is a measure of the flux radiated isotropically per unit volume of the source per unit solid angle in a particular direction, and is $L^* = d^2\Phi(dv\,d\omega)$ (flux per cubic meter steradian $= L/$meter).

Unfortunately the entity of sterisent (flux per cubic meter steradian) is not appropriate for certain calculations involving excitation states in the emission and scattering of atmospheric gases and the phluometry scheme has not been worked out to include a term and symbol that includes the important volume emission rate (quanta per second cubic meter). We therefore use the general term *volume emission rate* σ (flux per cubic meter), which is the geometrical distribution of flux with respect to position in a volume. We recommend the terms *radiant volume emission rate* σ_e (watts per cubic meter), *luminous volume emission rate* σ_v (lumens per cubic meter), and *photon volume emission rate* σ_p (quanta per second cubic meter).

The average *volume emission rate* σ_{av} of a source is the ratio of the flux rate to the volume. The limiting value of the average volume emission rate as the volume is reduced is the *volume emission rate* σ. The volume emission rate is a measure of the flux radiated isotropically per unit volume of the source at a point and is $\sigma = d\Phi/dv$ (flux per cubic meter).

The utility of the phluometry scheme, where a general term is devised to represent a geometrical distribution of flux which can be extended to other forms of flux through the use of appropriate modifiers, is evident in the preceding case of volume emission rate.

There are other entities [12] not listed in Tables 3-2 or 3-3 that fit in the phluometry scheme, such as *exposure H*, *fluence F*, and *fluence rate F_t* that deal with applications in photobiology or photochemistry; however, they shall not be considered here.

3-4   PHOTOMETRIC ENTITIES

Generally speaking, photometry is the study of the transfer of radiant energy in the form of light. Rather than measuring light in terms of energy rate, the visual effect is taken into account. The visual sensitivity of the "standard observer" is described in terms of the response of the human eye.

There exists a direct correspondence among photometric and radiometric units as shown in Tables 3-1 and 3-2 which list the basic entities.

Unfortunately, there is considerable overabundance of terminology being used in the field of photometry. The following terminology is in use [5].

The term *candle power* is used as an alternative to *luminous intensity* and has the units of *candelas* (formerly candle), which is equivalent to lumens per unit solid angle.

The term *illumination* is used as an alternative to *illuminance* and has the units of *footcandle* (lumens per square foot), *lux* (lumens per square meter), and *phot* (lumens per square centimeter).

The term *photometric brightness* is used as an alternative to *luminance* and has the units of candela per unit area, *stib* (candela per square centimeter), *nit* (candela per square meter), *foot lambert* ($1/\pi$ candela per square foot), *lambert* ($1/\pi$ candela per square centimeter), *apostilb* ($1/\pi$ candela per square meter).

The term *luminous efficacy k* has the units of lumens per watt. NBS has recommended that until the value of k is determined with an uncertainty no greater than 1%, the value of $k = 680$ lm W^{-1} be used [13].

3-5   PROJECTED AREA

There are additional entities, not given in Tables 3-2 or 3-3 that are necessary in order to calculate the geometrical transfer of flux. They are listed in Table 3-4; all are defined in the succeeding sections.

TABLE 3-4

Additional Radiometric Entities

Terms	Symbols	Units
Projected area	$A_p = \int_A \cos\theta\, dA$	m^2
Solid angle	$\omega = \int_0^{2\pi} d\phi \int_0^\Theta \sin\theta\, d\theta = 2\pi(1 - \cos\Theta)$	sr
Projected solid angle	$\Omega = \int_\omega \cos\theta\, d\omega = \pi\sin^2\Theta$	sr
Relative aperture[a]	$(f\text{-no}) = (2\pi \sin\Theta)^{-1}$	unitless
Throughput	$T = A \int_\omega \cos\theta\, d\omega = A\Omega$	sr m^2
Absorptance	α	unitless
Transmittance	τ	unitless
Reflectance	ρ	unitless
Emissivity (emittance)	ε	unitless

[a] (f-no) stands for f-number.

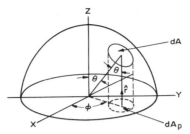

Figure 3-1   Illustration of projected area.

The area of a rectilinear projection of a surface (not necessarily a plane surface) onto a plane perpendicular to the unit vector $\hat{r}$ is the *projected area* A_p and is given by

$$A_p = \int_A \cos\theta \, dA \quad [\text{m}^2], \tag{3-1}$$

where θ is the polar angle between the unit vector $\hat{r}$ and the normal to the element dA of the surface. This can be visualized in terms of an element of the surface of a hemisphere projected onto the base of the hemisphere as illustrated in Fig. 3-1.

3-6   SOLID ANGLE

The *solid angle* ω of a cone formed by straight lines from a single point (the vertex) can be defined as the area intercepted on the surface of a unit hemisphere (radius equal to unity) centered at the vertex (see Fig. 3-2). The element of solid angle in spherical coordinates on the unit sphere is given by

$$d\omega = dA/r^2 = \sin\theta \, d\theta \, d\phi \quad [\text{sr}]. \tag{3-2}$$

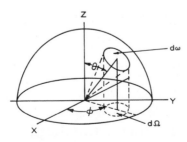

Figure 3-2   Illustration of solid angle and projected solid angle.

The solid angle for a right circular cone oriented with its center on the Z-axis is given by

$$\omega = \int_0^{2\pi} d\phi \int_0^{\Theta} \sin\theta \, d\theta = 2\pi(1 - \cos\Theta) \quad [\text{sr}], \tag{3-3}$$

where Θ is the half-angle of the solid angle. This is illustrated in Fig. 3-3.

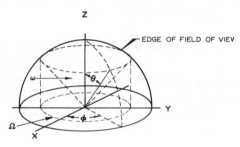

Figure 3-3   Illustration of solid angle as a right circular cone with its center aligned along the Z-axis.

3-7   PROJECTED SOLID ANGLE

The projected solid angle element is given by

$$d\Omega = \cos\theta \, d\omega = \cos\theta \sin\theta \, d\theta \, d\phi \quad [\text{sr}], \tag{3-4}$$

where θ is the angle between the cone axis and the zenith axis of the unit hemisphere (see Fig. 3-2), and may be visualized as the projection of the area (solid angle $d\omega$) of the unit hemisphere onto the base of the hemisphere. The *projected solid angle* Ω for a right circular cone oriented with its center on the Z-axis is given by

$$\Omega = \int_0^{2\pi} d\phi \int_0^{\Theta} \sin\theta \cos\theta \, d\theta = \pi \sin^2\Theta \quad [\text{sr}], \tag{3-5}$$

where Θ is the half-angle edge of the solid angle (see Fig. 3-3). The value of Ω is equal to ω [Eq. (3-3)] to within 1% for angles Θ less than 10°.

For a full hemisphere ($\theta = 90°$), Eq. (3-3) yields 2π sr, which is one-half the area of a unit sphere, while Eq. (3-5) yields π, which is the area of the base of a unit hemisphere.

3-8 THROUGHPUT AND f-NUMBER

The *basic throughput, optical extent,* or *étendue* [14–16] is the product of the index of refraction n, the area A, and the projected solid angle Ω. Because the basic throughput is invariant, it applies to the entrance pupil of a sensor, to detectors and sources, and can be applied to a beam of radiation in the abstract. The product $A\Omega$ is known as the *throughput* or *geometrical extent* and is represented by the symbol T. The throughput is also invariant in a homogeneous environment. Thus

$$T = A\Omega = A\pi \sin^2 \Theta \quad [\text{sr m}^2]. \tag{3-6}$$

The *relative aperture* or f-number is another measure of the "light gathering power" of an optical system which is defined as the ratio of the effective focal length to the diameter. The symbol for f-number is $(f\text{-no})$ and has a value

$$(f\text{-no}) = (2n \sin \alpha)^{-1} \quad [\text{unitless}], \tag{3-7}$$

where α is defined in Fig. 3-4 and the quantity $n \sin \alpha$ is referred to as *numerical aperture*. The relationship between solid angle and f-number is given approximately as

$$\Omega = \pi \sin^2 \alpha = \tfrac{1}{4}\pi(f\text{-no})^{-2} \quad [\text{sr}]. \tag{3-8}$$

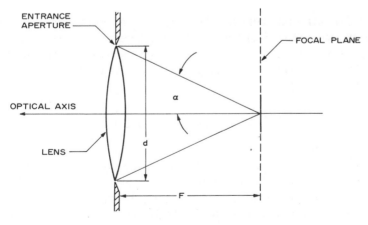

Figure 3-4 The relationship between f-number and the cone angle α.

3-9   PROPERTIES OF MATERIALS

Terms that end in -*ivity* refer to the ideal property of a material having planar surfaces between two media and no oxides or coatings on the surface. Terms ending in -*ance* refer to the property of an actual sample or path.

The *absorptance* α is the ratio of the power absorbed by a surface to the total incident power.

The *transmittance* τ is the ratio of the power transmitted through a surface to the total incident power.

The *reflectance* ρ is the ratio of the power reflected off a surface to the total incident power.

Thus we speak of the transmittance of a filter, the reflectance of a mirror, or the absorptance of a sample.

The *emissivity* ε is the ratio of the power radiated by the surface of a body to that of an ideal *blackbody* radiator. The value of ε for a blackbody is unity.

3-10   THE RAYLEIGH

An extended source is best qualified in terms of the radiant sterance [radiance] L_e (watts per square meter steradian) or photon sterance L_p (quanta per second square meter steradian). The units of photon sterance are sometimes modified (by a constant) for convenience in making atmospheric studies. The resulting photon sterance $L_p{}'$ is given in units of *rayleighs* (R),† and is obtained by multiplying the photon sterance L_p by the factor $4\pi \times 10^{-10}$. Thus a rayleigh may be viewed as $1/(4\pi) \times 10^{10}$ q sec^{-1} m^{-2} sr^{-1}.

The Baker [17] formula can be used to convert from radiant sterance L (watts per square meter steradian) to the rayleigh

$$L_p{}' = 2\pi\lambda L_e \times 10^9 \quad [\text{R}]. \tag{3-9}$$

Equation (3-9) is derived as follows: Chamberlain [18] defines the photon sterance I as

$$I = 10^{-10}L_p \quad [\text{q sec}^{-1} \text{ m}^{-2} \text{ sr}^{-1}] \tag{3-10}$$

while the rayleigh is given by definition as

$$L_p{}' = 4\pi I = 4\pi \times 10^{-10}L_p \quad [\text{R}]. \tag{3-11}$$

---

† This symbol R, used for rayleigh, is not to be confused with R responsivity as used in later sections of this book.

The energy of a photon of wavelength λ (given in micrometers) is

$$\mathscr{E}(\lambda) = (hc/\lambda) \times 10^6 \quad [\text{J}], \tag{3-12}$$

where h is Planck's constant and c the speed of light in a vacuum. Therefore, the radiant sterance [radiance] L_e is, from Eqs. (3-11) and (3-12),

$$L_e = L_p(hc/\lambda) \times 10^6 = L_p'(hc/4\pi\lambda) \times 10^{16} \quad [\text{W m}^{-2}\,\text{sr}^{-1}]. \tag{3-13}$$

Using the values for the atomic constants given in Appendix C,

$$L_e \cong L_p' \times 10^{-9}/2\pi\lambda \quad [\text{W m}^{-2}\,\text{sr}^{-1}] \tag{3-14}$$

to within 1%. Solving for L_p' yields Eq. (3-9).

3-11   SPECTRAL RADIOMETRIC ENTITIES

The radiometric entities of areance, sterance, pointance, volume emission rate, and sterisent are entities that are differential with respect to area, solid angle, and volume, and must be integrated over the appropriate variable to obtain the total flux.

There are also entities defined to be differential with respect to wavelength or optical frequency [11, p. 1433]. One of these is the *spectral radiant power* $P(\lambda) = dP/d\lambda$ which is the ratio of the radiant power to the wavelength interval $\Delta\lambda$, as $\Delta\lambda$ is reduced to a particular wavelength, and has the units of watts per unit wavelength. The value of $P(\lambda)$, in general, may vary with wavelength. The total radiant power between the wavelengths λ_1 and λ_2 is given by

$$P = \int_{\lambda_1}^{\lambda_2} P(\lambda)\, d\lambda.$$

Other spectral quantities are *spectral areance* $E(\lambda) = M(\lambda) = d^2\Phi/dA\ d\lambda$, *spectral pointance* $I(\lambda) = d^2\Phi/d\Omega\ d\lambda$, *spectral sterance* $L(\lambda) = d^3\Phi/dA\ d\Omega\ d\lambda$, *spectral sterisent* $L(\lambda)^* = d^3P/dv\ d\Omega\ d\lambda$ and *spectral rayleigh radiance* $R(\lambda)$. These spectral density functions may be given in terms of frequency (wave number) increment, for instance $L(\bar{v}) = d^3\Phi/dA\ d\Omega\ d\bar{v}$.

3-12   APPARENT RADIOMETRIC ENTITIES

The calculations of the radiometric characteristics of an observed source in an attenuating medium from measurements made at a distance always involve assumptions about the nature of the absorptions, emissions, and scattering of the radiation within the intervening medium [11, p. 1433].

The effect of the intervening medium is to introduce errors in the calculated values unless corrections are made for these effects. In the absence of such corrections, the calculated values are called "apparent" quantities. The radiant pointance [intensity] of a star, calculated from measurements made on the earth's surface with no correction for the atmospheric attenuation, would be the *apparent radiant pointance* of the star. On the other hand, calculations of the total overhead radiant sterance [radiance] of the sky are defined to include the effects of emission, absorption, and scattering, and therefore extend from the radiometer outward. In this case, the word "apparent" would not apply.

The incident areance [irradiance] at the aperture of the radiometer is by definition the flux density falling on the radiometer aperture; therefore, the word "apparent" is never appropriate.

REFERENCES

1   F. E. Nicodemus, Optical resource letter on radiometry. *Am. J. Phys.* **38**, 43–49 (1970).
2   "International Lighting Vocabulary," 3rd Ed., Publ. CIE No. 17 (E-1.1), common to the CIE and IEC, International Electrotechnical Commission (IEC), International Commission on Illumination (CIE). Bur. Cent. CIE, Paris, 1970.
3   C. H. Page and P. Vigoureux, eds., The International System of Units (SI). *Natl. Bur. Stand. (U.S.), Spec. Publ.* No. 330 (1974).
4   Optical Society of American Nomenclature Committee Report, *J. Opt. Soc. Am.* **57**, 854 (1967); see also review of ref. 5, *J. Opt. Soc. Am.* **58**, 864 (1968).
5   "American National Standard Nomenclature and Definitions for Illuminating Engineering," RP-16 (ANSI Z7.1-1967). Illum. Eng. Soc., New York, 1967.
6   F. E. Nicodemus, "Proposed Military Standard—Infrared Terms and Definitions," Part 1 of 2 Parts, p. 3.1-9. Am. Soc. Mech. Eng., New York, 1971.
7   I. J. Spiro, Radiometry and photometry. *Opt. Eng.* **13** (1974); **14** (1975); **15** (1976). Contains a column in which proposals are aired in each issue; see esp. **13**, G183–G187 (1974); **15**, SR-7 (1976).
8   R. C. Jones, Terminology in photometry and radiometry. *J. Opt. Soc. Am.* **53**, 1314 (1963).
9   J. R. Meyer-Arendt, Radiometry and photometry: Units and conversion factors. *Appl. Opt.* **7**, 2081–2084 (1968).
10 J. Geist and E. Zalewski, Chinese restaurant nomenclature for radiometry. *Appl. Opt.* **12**, 435 (1973).
11 F. E. Bell, Radiometric quantities, symbols, and units. *Proc. IRE* **47**, 1432–1434 (1959).
12 F. E. Nicodemus, "Reference Book on Radiometric Nomenclature." U.S. Nav. Weapons Cent., China Lake, California (draft copy, personal communications) (1974).
13 National Bureau of Standards, U.S. Department of Commerce, *Opt. Radiat. News* No. 7, p. 2 (1975).
14 F. E. Nicodemus, Radiometry. *In* "Optical Instruments," Part 1 (R. Kingslake, ed.), Applied Optics and Optical Engineering, Vol. 4, Ch. 8. Academic Press, New York, 1967.
15 W. H. Steel, Luminosity, throughput, or etendue? *Appl. Opt.* **13**, 704 (1974).
16 F. E. Nicodemus, Radiance. *Am. J. Phys.* **31**, 368 (1963).
17 D. J. Baker, Rayleigh, the unit for light radiance. *Appl. Opt.* **13**, 2160–2163 (1974).
18 J. W. Chamberlain, "Physics of the Aurora and Airglow," p. 569. Academic Press, New York, 1961.

CHAPTER

IV

Blackbody Radiation

4-1 INTRODUCTION

All objects that have a temperature at any value other than absolute zero are continuously emitting and absorbing radiation. The radiation characteristics of certain surfaces are completely specified if the temperature is known. These surfaces radiate continuously through the optical spectrum, and are known as "ideal thermal radiators" or "blackbodies" [1–3]. There are blackbody simulators commercially available that constitute a good approximation to blackbody radiation over a useful wavelength region.

4-2 PLANCK'S EQUATION

Blackbody simulators are used as primary calibration standards. A practical cavity source approaches ideal blackbody radiation. The accuracy of such a laboratory standard is primarily determined by the accuracy with which its temperature can be determined. Large area blackbody simulators have additional problems of reduced emissivity and nonuniform temperature distribution. Blackbody radiation is described by *Planck's equation*†° where the values of the constants are given in Appendix C.

---

† This relation was formulated by Max Karl Ernst Ludwig Planck (1858–1947), the German Nobel prize-winning physicist who originated the quantum theory.

The spectral sterance [radiance] as a function of absolute temperature and wavelength is given as

$$L(\lambda) = (2hc^2/\lambda^5)(\exp(hc/\lambda kT) - 1)^{-1} \quad [\text{W m}^{-3}\text{ sr}^{-1}], \qquad (4\text{-}1)$$

where h is Planck's constant $= 6.6262 \times 10^{-34}$ J sec, c the velocity of light $= 2.9979 \times 10^8$ m sec, λ the wavelength (m $= 10^6$ μm), k Boltzmann's constant† $= 1.3806 \times 10^{-23}$ J K^{-1}, and T absolute temperature (kelvin).

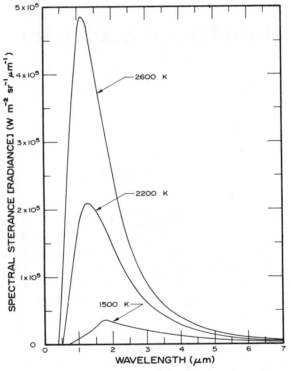

Figure 4-1 $L(\lambda)$ as a function of wavelength and temperature.

The unit for sterance [radiance] is given in watts per cubic meter steradian in Eq. (4-1) where meter (m) is used as the unit of wavelength rather than the micrometer (μm). This should not be confused with volume concentration, which it is not.

† The constant is named after Ludwig Edvard Boltzmann (1844–1906), the Austrian physicist who demonstrated the Stefan–Boltzmann law on radiation from a blackbody.

A convenient form of the equation that may be evaluated with a desk calculator is

$$L(\lambda) = (1.191066 \times 10^8/\lambda^5)$$

$$(\exp(1.43883 \times 10^4/\lambda T) - 1)^{-1} \quad [\text{W m}^{-2} \text{ sr}^{-1} \ \mu\text{m}^{-1}] \qquad (4\text{-}2)$$

where λ is entered directly in micrometers (μm).

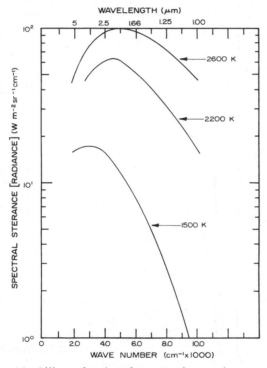

Figure 4-2 $L(\lambda)$ as a function of wave number $\bar{\nu}$ and temperature.

Figures 4-1, 4-2, and 4-3 illustrate the spectral distribution of blackbody radiation. Figure 4-1 is a linear plot with wavelength as abscissa and spectral radiant sterance [radiance] $L(\lambda)$ as ordinate, which illustrates the following:

(1)   The spectral radiant sterance [radiance] increases at all wavelengths for increased temperatures.
(2)   The peak of the curve shifts toward shorter wavelengths for higher temperatures.

(3)   The ratio $\Delta L(\lambda)/\Delta T$ has its greatest value in those regions of the curve where the wavelength is less than the wavelength of the peak radiation, that is, the region to the left side of the peak.

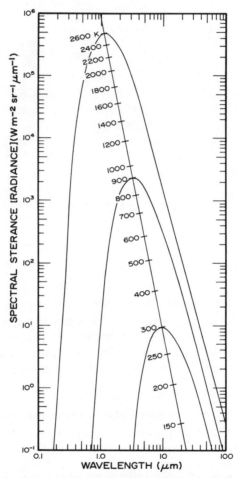

Figure 4-3   Log–log plot of $L(\lambda)$ as a function of wavelength and temperature.

Figure 4-2 is also a linear plot, but with wave number as abscissa. It illustrates the different perspective of blackbody radiation in terms of wave number from which it is observed that the maximum occurs at a different wavelength for $L(\bar{\nu})$.

Figure 4-3 is a log–log plot of blackbody radiation which illustrates the following:

(1) The shape of the blackbody radiation curve is exactly the same for any temperature T.

(2) A curve connecting the peak radiation for each temperature is a straight line.

(3) The shape of the curve can be shifted along the straight line connecting the peaks to obtain the curve at any temperature.

A "do-it-yourself" slide rule can be constructed by placing a sheet of tracing paper over Fig. 4-3 and tracing the curve and the line. Then, by keeping the lines overlapping and setting the peak at the desired temperature, the tracing becomes the blackbody curve for that temperature.

Tables have been published [4] that contain compilations of various blackbody functions including the solution to Eq. (4-1) for spectral radiant sterance [radiance]. The tables are especially useful for precision work; however, interpolation is necessary for intermediate values.

Slide rules are also available that provide for rapid calculation of blackbody quantities with good accuracy.

4-3 THE WIEN DISPLACEMENT LAW

The Planck radiation formula shows that the spectrum of the radiation shifts toward shorter wavelengths as the temperature of the radiator is increased. The derivative of the Planck equation [Eq. (4-1)] with respect to wavelength yields the *Wien displacement law*† which gives the wavelength for which maximum radiation occurs for a given temperature [5].

$$\lambda_m = 2898/T \quad [\mu m], \tag{4-3}$$

The solution to this equation provides for the wavelength designation along the straight line shown in Fig. 4-3.

The wavelength λ_m at which the peak or maximum radiation occurs is significant for calibration purposes. At wavelengths less than λ_m the spectral power density changes very rapidly with temperature and wavelength, so that a slight error in determining either the temperature or the wavelengths results in a relatively large error in the radiation. However, at wavelengths beyond λ_m, the radiation changes less rapidly with wavelength and is less sensitive to temperature errors as is evident in Fig. 4-1.

† This relation is named after Wilhelm Wien (1864–1928), the German physicist who received the Nobel prize for its discovery.

Another factor that deserves attention is the uniformity (or lack of uniformity) of the radiation as a function of wavelength. It would be desirable to make use of a standard source with a uniform spectrum for the calibration of a spectrometer. This would result in uniform stimulation of the spectrometer response at all wavelengths throughout its free spectral range. However, the blackbody curve is very nonuniform except in the region of wavelengths near λ_m.

For these reasons it is desirable to use relatively hot blackbody temperature calibration sources where the wavelengths of interest are at, or beyond, the wavelength λ_m of maximum radiation. However, this calibration ideal is often difficult to follow because of the problem of operating and maintaining high-temperature blackbodies (that is, blackbodies that operate above 1000°C where the materials begin to glow red-hot and suffer oxidation). Also, such high-temperature blackbody sources tend to overdrive or saturate sensitive electrooptical sensors.

For example, the calibration of a uv photometer that is designed to measure the 3940-Å (0.3940-μm) molecular nitrogen ion first negative band [6] would, by Eq. (4-3), require a blackbody operated at a temperature greater than 7000K to conform to the ideal previously outlined. Tungsten lamps are often used as standard sources at temperatures above 1000°C.

4-4 THE STEFAN–BOLTZMANN LAW

The total power radiated per unit area of a blackbody is obtained by integrating Planck's radiation law over all wavelengths, and is known as the *Stefan–Boltzmann law*† [5, p. 15]

Radiant areance [radiant exitance] $M = \varepsilon\sigma T^4$ [W m^{-2}],   (4-4)

where ε may be called the "total hemispherical emissivity." For many purposes ε may be taken as a constant in the case of a solid, and is characteristic of the solid surface. The *Stefan–Boltzmann constant* σ has a value of 5.66961×10^{-8} and has the units of watts per square meter kelvin^4. At room temperature (approximately 300K), a perfect blackbody ($\varepsilon = 1$) of area equal to 1 m^2 emits a total power of 460 W. If its surroundings are at the same temperature, it absorbs the same amount.

The heat loss from a blackbody at temperature T_1 to its surroundings which are at temperature T_2 is given by

$$M = \varepsilon\sigma(T_1{}^4 - T_2{}^4)   [\text{W m}^{-2}].   (4-5)$$

---

† Josef Stefan (1835–1893), Austrian physicist, originated this law.

4-5   RAYLEIGH–JEANS' LAW AND WIEN'S RADIATION LAW

Two well-known historical approximations [7] to Planck's law are readily obtained from Eq. (4-1). For the conditions $hc/\lambda kT \ll 1$, Eq. (4-1) reduces to

$$L(\lambda) \simeq 2ckT/\lambda^4 \quad [\text{W m}^{-3}\text{ sr}^{-1}] \tag{4-6}$$

which is known as *Rayleigh–Jeans' law*,† valid only at long wavelengths.

For the conditions $hc/\lambda kT \gg 1$, Eq. (4-1) reduces to

$$L(\lambda) \simeq (2hc^2/\lambda^5)\exp(-hc/\lambda kT) \quad [\text{W m}^{-3}\text{ sr}^{-1}] \tag{4-7}$$

which is known as *Wien's radiation law*, valid only at short wavelengths.

4-6   EMISSIVITY AND KIRCHHOFF'S LAW

As previously indicated, practical sources approach the ideal blackbody. The ideal receiver and radiator of radiant energy is called a *blackbody radiator* for reasons that shall be discussed next.

If a small solid object S is located within an evacuated isothermal cavity, according to the second law of thermodynamics there will be a net flow of heat between the object and the walls toward the cooler of the two. Eventually, the object will come to equilibrium temperature with the cavity walls, and remain at that temperature. If the object absorbs only a portion α of the incident areance [irradiance] E (watts per square meter) which a perfectly black, ideal radiator would emit at that temperature, then the object S will emit an amount εM equal to that absorbed; that is $\alpha E = \varepsilon M$. This is an expression of *Kirchoff's law*,‡ which states that the absorptivity α of a surface is exactly equal to the emissivity ε of that surface [5, p. 14].

The relationship

$$P_i = P_\alpha + P_\rho + P_\tau \quad [\text{W}], \tag{4-8}$$

† This relation is named after the English mathematical physicists John William Strutt Rayleigh, 3rd Baron, 1842–1919 who suggested the notion and Sir James Hopwood Jeans (1877–1946) who arrived at the formula.

‡ This law is named after Gustav-Robert Kirchhoff (1824–1887), German physicist, who with Bunsen discovered spectrum and analysis and who formulated Kirchhoff's laws of electricity.

where P_i is the incident power, is a statement of the conservation of energy. Dividing both sides by P_i yields

$$\alpha + \rho + \tau = 1 \quad \text{[unitless]}. \tag{4-9}$$

The terms $\varepsilon, \alpha, \rho$, and τ are defined in Table 3-4. For an opaque body, $\tau = 0$, so Eq. (4-9) becomes

$$\alpha = 1 - \rho, \tag{4-10}$$

indicating that the surfaces of high reflectance are poor emitters. That is the reason why the ideal emitter is literally a diffuse black surface, or a blackbody.

Generally the emissivity ε of a surface is a function of wavelength, temperature, and direction. For many cases of radiation in solids ε can be considered constant. A radiating body is known as a *greybody* when $\varepsilon < 1$, and is independent of λ. The power spectral density curve of a greybody has the same shape as that for a blackbody, but at any wavelength it has a value that bears the ratio ε to that of an ideal blackbody.

4-7   LAMBERT'S COSINE LAW

The radiation per unit solid angle (in a specific direction) from a flat zero-thickness surface element varies with the angle made with the normal to the surface. It is easily deduced that the amount of energy in a given solid angle varies in proportion to the cosine of the angle between the direction in question and the normal to the surface. This is known as *Lambert's cosine law*† [5, p. 42] (see Fig. 4-4), and explains why a spherical body with relatively uniform surface conditions, such as the sun, appears as a flat disk. The

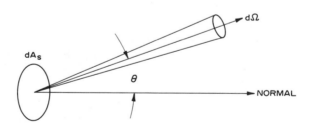

Figure 4-4   Radiation in a direction θ (with respect to the normal).

---

† This law bears the name of Johann Heinrich Lambert (1728–1777), the German mathematician, astronomer, and physicist who espoused the principles of light reflected and scattered from surfaces. This bright and versatile man was also the one who introduced the hyperbolic functions sinh and cosh, and he also proved that π is irrational.

projected area near the limb represents more real area, but that area is radiating less in the oblique direction so the effects cancel and the disk looks uniformly bright. Lambert's law can be expressed in terms of radiant pointance [intensity]

$$I = I_0 \cos \theta \quad [\text{W sr}^{-1}]. \tag{4-11}$$

When the geometrical distribution of radiation from a surface obeys Lambert's cosine law, the source is said to be "Lambertian" or perfectly diffuse.

REFERENCES

1   J. C. DeVos, Evaluation of the quality of a blackbody. *Physica (Utrecht)* **20**, 669 (1954).
2   W. L. Eisenman and A. J. Cussen, A comparative study of several black bodies. *Proc. IRIS* **1**, 39 (1956).
3   C. S. Williams, Discussion of the theories of cavity-type sources of radiant energy. *J. Opt. Soc. Am.* **51**, 564 (1961).
4   M. Pivovonsky and M. R. Nagel, "Tables of Blackbody Radiation Functions." Macmillan, New York, 1961.
5   P. W. Kruse, L. D. McGlauchlin, and R. B. McQuistan, "Elements of Infrared Technology," p. 29. Wiley, New York, 1962.
6   R. C. Whitten and I. G. Poppoff, "Fundamentals of Aeronomy," p. 194. Wiley, New York, 1971.
7   R. A. Smith, F. E. Jones, and R. P. Chasmar, "The Detection and Measurement of Infra-Red Radiation," pp. 25–35. Oxford Univ. Press, London and New York, 1957.

CHAPTER

V

Geometrical Flux Transfer

5-1 INTRODUCTION

The calculation of the transfer of flux from a source to a reference surface can be difficult until a familiarity with the properties of geometrical radiation is acquired. Dimensional (unit) analysis provides not only a check on the correctness of a calculation but also leads to an understanding of the basic geometry involved in radiation transfer.

The utility of the consideration of the radiometric entities as field entities was indicated in an earlier chapter. This concept shall be expanded in this chapter in terms of sterance [radiance, luminance] L (flux per square meter steradian) to illustrate the meaning of a ray (or beam), of optical throughput, and of the invariance of the quantity L/n^2, where n is the index of refraction of the medium.

Sample calculations are given in this chapter to illustrate flux transfer and dimensional analysis. These calculations demonstrate that the geometry of radiation is axiomatic and that the entire theory can be deduced from the definition of sterance [radiance, luminance].

5-2   THE RAY

A *ray* is defined [1] as a beam of flux (light or other radiant energy) of small or infinitesimal cross section. Such a ray, shown in Fig. 5-1, consists of a beam of flux which is bounded at two nonzero area points of cross-sectional area ΔA_1 and ΔA_2 which are separated by the distance s. All the flux in the beam that passes through element ΔA_1 must also pass through element ΔA_2.

5-3   THE INVARIANCE OF THROUGHPUT

The projected area of each cross-sectional area element shown in Fig. 5-1 in the direction of s is given by $\Delta A_1 \cos \theta$ and $\Delta A_2 \cos \phi$ where N is the normal to the cross-sectional area element. The linear dimensions of ΔA_1

Figure 5-1   Geometry of a ray.

and ΔA_2 (which are not necessarily the same) are assumed to be very small compared with s, so that the solid angle subtended at any point on ΔA is uniquely given by the projected area of the opposite element divided by s^2. Thus the product of the projected area and the solid angle subtended at the area for each end is given by

$$\Delta T_1 = \Delta A_1 \cos \theta \, \Delta\omega_1 = \Delta A_1 \cos \theta (\Delta A_2 \cos \phi)/s^2$$

$$= \Delta A_1 \, \Delta\Omega_1 \quad [\text{sr m}^2], \tag{5-1}$$

$$\Delta T_2 = \Delta A_2 \cos \phi \, \Delta\omega_2 = \Delta A_2 \cos \phi (\Delta A_1 \cos \theta)/s^2$$

$$= \Delta A_2 \, \Delta\Omega_2 \quad [\text{sr m}^2], \tag{5-2}$$

where $\cos \theta \, \Delta\omega_1 = \Delta\Omega_1$, $\cos \phi \, \Delta\omega_2 = \Delta\Omega_2$, and $\Delta\Omega$ the projected solid angle. This product is the incremental throughput ΔT. The quantity $\Delta T_1 = \Delta T_2$, as given by Eqs. (5-1) and (5-2) which indicates the invariance of throughput in a homogeneous medium. The product nT (the basic throughput) is invariant in a nonhomogeneous medium [2–4].

5-4 THE INVARIANCE OF STERANCE [RADIANCE, LUMINANCE]

By the definition of a ray, the flux flowing along the beam is invariant; that is, the flux passing through ΔA_1 of Fig. 5-1 is equal to that passing through ΔA_2. The throughput of a ray is also constant, as was just shown. From these two facts, it can be deduced that the sterance is also constant along a ray [4, p. 373]. This is accomplished by taking the ratio of the flux passing through the elemental area to the product of the projected area in the direction s and the solid angle into which the flux is radiating. By definition, this ratio, taken in the limit, is the sterance L of the area ΔA, as given by

$$\Delta\Phi_1/\Delta T_1 = \Delta\Phi_1/(\Delta A_1 \cos\theta \; \Delta\omega_1) \simeq L_1, \qquad (5\text{-}3)$$

$$\Delta\Phi_2/\Delta T_2 = \Delta\Phi_2/(\Delta A_2 \cos\phi \; \Delta\omega_2) \simeq L_2. \qquad (5\text{-}4)$$

The quantity $L_1 = L_2$, as indicated in Eqs. (5-3) and (5-4). The utility of the consideration of sterance L as a field entity becomes apparent when areas ΔA_1 and ΔA_2 are taken to be any convenient surface such as the area of a source or the aperture of an electrooptical sensor.

Sterance has the same geometrical characteristics as the photometric entity of "brightness" [luminance]. When a diffuse (nonreflective) surface that obeys Lambert's law is observed by the eye, the apparent brightness is independent of the direction or of the distance from which it is observed. This also explains why the photographic exposure of a distant sunlit landscape is usually independent of distance or direction.

5-5 THE FUNDAMENTAL THEOREM OF RADIOMETRY

The analysis of the previous section applies to a ray in an isotropic medium and shows that the value of the sterance along a ray is invariant. This result may be extended to an important generalization known as the "brightness theorem" [5]. This theorem may be given as the invariance of the quantity L/n^2, where n is the index of refraction [4, p. 373; 6]. An equivalent statement pertains to the apparent brightness of a source viewed through any specular optical system (any system containing only elements that specularly transmit or reflect, and for which there are no losses due to reflection, absorption, scattering, or refraction). This theorem states that if the sterance of a source in the direction of the optical axis is L_1, and if the refractive indices of the medium in which the source and observer are located are n_1 and n_2, respectively, then the sterance L_2 of the source viewed through the optical system is

$$L_2 = L_1(n_2/n_1)^2 \quad [\Phi \text{ m}^{-2} \text{ sr}^{-1}]. \tag{5-5}$$

This is true regardless of the diameter of the aperture or the number of components in the system. In most cases, the source and the observer are in the same medium so that $n_1 = n_2$; thus this theorem implies that it is impossible to increase the apparent brightness of a source with any type of an optical system.

An exception to this conclusion, which is an application of the theorem, applies to the special case of "immersion optics." The apparent brightness of a ray is increased by placing the reference surface within a medium of higher index of refraction. A detector–transducer can be immersed in a collector lens of higher than air refractive index to obtain a higher detectivity [7].

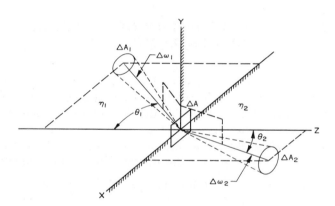

Figure 5-2   A refracted ray.

The proof of the brightness theorem is as follows: The beam depicted in Figure 5-1 is permitted to pass through a refractive surface from a medium of index n_1 into a medium of index n_2 as shown in Fig. 5-2 (the angle ϕ of the ray is not altered by refraction since the ray remains in the XZ-plane). The flux incident upon the refractive surface from ΔA_1 is given by

$$\Delta \Phi = L_1(\Delta A_1 \, \Delta A \cos \theta_1)/s^2 = L_1 \, \Delta T_1 \quad [\Phi]. \tag{5-6}$$

The same flux leaves ΔA within the angle $\Delta \omega_2$ so the apparent sterance L_2 of ΔA is

$$L_2 = \Delta \Phi / \Delta T_2 = L_1 \, \Delta T_1 / \Delta T_2$$

$$= L_1 \, (\Delta \omega_1 \, \Delta A \cos \theta_1 / \Delta \omega_2 \, \Delta A \cos \theta_2) \quad [\Phi \text{ m}^{-2} \text{ sr}^{-1}]. \tag{5-7}$$

Snell's law of refraction governs the relationship between θ_1 and θ_2 which is

$$n_1 \sin \theta_1 = n_2 \sin \theta_2. \tag{5-8}$$

By differentiation, we obtain

$$n_1 \cos \theta_1 \, d\theta_1 = n_2 \cos \theta_2 \, d\theta_2. \tag{5-9}$$

Multiplying Eq. (5-8) by Eq. (5-9) and changing to incrementals, we have

$$(n_2/n_1)^2 = \sin \theta_1 \, \Delta\theta_1/(\sin \theta_2 \, \Delta\theta_2)(\cos \theta_1/\cos \theta_2)$$
$$= (\Delta\omega_1 \cos \theta_1)/(\Delta\omega_2 \cos \theta_2). \tag{5-10}$$

Substitution of Eq. (5-10) into (5-7) yields the desired results:

$$L_2 = L_1(n_2/n_1)^2 \quad [\Phi \text{ m}^{-2} \text{ sr}^{-1}]. \tag{5-11}$$

5-6   THE BASIC ENTITY OF STERANCE [RADIANCE, LUMINANCE]

The greatest degree of generality is achieved by specifying the geometrical distribution of flux with respect to all the parameters, that is, both position and direction, and to relate that distribution to the total flux in the beam or ray. This is accomplished with the distribution function of sterance [radiance, luminance] L which is defined as the flux per unit projected area and solid angle or the flux per unit throughput. The defining equation is

$$L = d^2\Phi/(d\omega \cos \theta \, dA) = d^2\Phi/(d\Omega \, dA) = d\Phi/dT \quad [\Phi \text{ m}^{-2} \text{ sr}^{-1}]. \tag{5-12}$$

The entire theory of radiation geometry or of radiation transfer (that is for propagation through an isotropic, passive, lossless medium) can be deduced from the definition of sterance.

The calculations given in this chapter to illustrate the transfer of radiation are based on geometrical or spatial concepts only. The real problem of calibration (as indicated in Chapter I) includes the domains of spectral, temporal, and polarization as well as spatial. However, for purposes of illustration, the variables can be considered as separable so that they can be considered one by one.

The calculations of the geometrical transfer of flux are based on the definition of sterance L (flux per square meter steradian). This is illustrated with reference to Fig. 5-3 as follows: Dimensional analysis indicates that the flux Φ, incident upon the collector, is obtained by taking the product of the source sterance L (flux per square meter steradian), the source projected

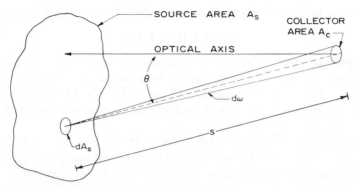

Figure 5-3   Geometry for the transfer of flux from the source area to the collector area.

area A_s (square meters), and the solid angle ω_s (steradians) into which the flux is radiating. The solid angle is defined by the collector aperture subtended at the source. The incremental flux is given by

$$\Delta\Phi = L \, \Delta A_s \cos \theta \, \Delta\omega_s. \tag{5-13}$$

In order to find the total flux it is necessary to integrate over both the area and the solid angle

$$\Phi = \int_{A_s} \int_{\omega_s} L \, dA_s \cos \theta \, d\omega_s. \tag{5-14}$$

Equation (5-14) can be written as

$$\Phi = L \int_{A_s} dA_s \int_{\omega_s} \cos \theta \, d\omega_s = LA_s\Omega_s \tag{5-15}$$

under certain conditions. In this case the solid angle Ω_s can be approximated by

$$\Omega_s = A_c/s^2, \tag{5-16}$$

where s is the distance between the source and A_c the collector area. Thus the flux is given approximately by

$$\Phi = LA_s A_c/s^2 = LT. \tag{5-17}$$

There is a "rule of thumb" in radiometry [8] that the relationship $A_s A_c/s^2 = T$ will yield accuracies of about 99% when the separation s is about 10 to 20 times the maximum transverse dimension of A_s or A_c. Thus Eq. (5-17) is often accurate enough for calibration. Then the invariance of throughput leads to the relationship

$$T = A_s\Omega_s = A_c\Omega_c, \tag{5-18}$$

where Ω_c is the solid angle of the source subtended at the collector A_c. Thus

$$\Phi = LA_s\Omega_s = LA_c\Omega_c. \qquad (5\text{-}19)$$

The right side of Eq. (5-19) may be interpreted in the case of an extended area source as follows: The flux in the beam is given by the product of the sterance L and the instrument throughput $A_c\Omega_c$. The left side of Eq. (5-19) applies for the case of a small source that does not completely fill the instrument field of view.

These considerations are illustrated for the case of the sun as a source. The various entities are calculated for a reference surface at the earth (above the attenuating atmosphere), assuming the sun to be a uniform isotropic radiator so that the average values correspond to the limiting values of the entities.

The radiant sterance [radiance] of the sun is the ratio of the total power P to the throughput of the sun, where

$$T_s = \int_A \cos\theta \, dA_s \int_\omega d\omega = A_p\omega, \qquad (5\text{-}20)$$

the projected area A_p is the area of the solar disk given by πr^2 (where r is the radius of the sun) and ω the solid angle of the sphere $(4\pi \text{ sr})$. Thus

$$L_s = P/(4\pi^2 r^2) = 1.98 \times 10^7 \quad [\text{W m}^{-2} \text{ sr}^{-1}] \qquad (5\text{-}21)$$

where $r = 6.96 \times 10^8$ m and $P = 3.79 \times 10^{26}$ W.

From Eq. (5-17) the power incident upon a *unit area* of surface at the earth is

$$P = L_s A_s A_c/s^2 = L_s \pi r^2/s^2 = 1340 \quad [\text{W}], \qquad (5\text{-}22)$$

where s the sun–earth distance is 1.5×10^{11} m and the reference area A_c is unity. This is identical to the radiant areance [irradiance] which has the units of watts per square meter and is known as the *solar constant*.

A sample calculation to illustrate the use of a blackbody calibration standard is as follows: An extended area blackbody is placed near the entrance aperture of a spectrometer to obtain a radiant sterance [radiance] calibration. The spectrometer records a 1.5-V output signal in response to a 500K blackbody temperature at 10 μm. The objective of the exercise is to find the sensor spectral responsivity $R(\lambda)$ which is the ratio of output voltage to extended area source spectral radiant sterance [radiance] $L(\lambda)$ at 10 μm.

The solution of Planck's equation for the blackbody source yields $L_b(\lambda) = 7.1 \times 10^1$ W m^{-2} sr^{-1} μm^{-1}. By Eq. (5-19) the spectral power incident upon the reference surface (the sensor entrance aperture) is given by the product of the blackbody sterance $L_b(\lambda)$ and the sensor throughput $A_c \Omega_c$ provided the source completely "fills" the instrument field of view Ω_c:

$$P(\lambda) = L(\lambda)A_c \Omega_c \quad [\text{W } \mu\text{m}^{-1}]. \qquad (5\text{-}23)$$

The sensor actually responds to the total power that is incident upon the detector after passing through the optical system. However, it can be calibrated in terms of the extended-area spectral radiant sterance [radiance] which is related to the spectral power by a constant. Another interpretation of the extended area source calibration is as follows: Based on the invariance of sterance, the spectral sterance at the reference surface (the sensor aperture) is equal to the source spectral sterance. In either case, the sensor spectral sterance responsivity is

$$R(\lambda) = V/L_b(\lambda) = 2.11 \times 10^{-2} \quad [\text{V}/(\text{W m}^{-2} \text{ sr}^{-1} \mu\text{m}^{-1})]. \qquad (5\text{-}24)$$

This example illustrates the utility of the consideration of radiant entities as field quantities. It is also significant that the calculation of the extended area source spectral sterance is independent of the sensor throughput in this case.

5-7   THE ENTITY OF POINTANCE [INTENSITY] AND POINT SOURCES

The pointance [intensity] can be defined for any source; however, it is not very meaningful except for sources that approximate a point source. It can be derived from the definition for the basic entity sterance [radiance, luminance] as follows. The flux incident upon a reference surface ΔA_2 from a source ΔA_1 (see Fig. 5-1) is given in terms of the sterance L by Eq. (5-14):

$$\Phi = \int_{\Delta\omega} \int_{\Delta A} L \, dA_1 \cos\theta \, d\omega_1 \quad [\Phi]. \qquad (5\text{-}25)$$

The pointance is given by definition as

$$\lim_{d\omega_1 \to 0} d\Phi/d\omega_1 = \int_{\Delta A} L \cos\theta \, dA_1 = L \cos\theta \, \Delta A_1 \quad [\Phi \text{ sr}^{-1}], \qquad (5\text{-}26)$$

where $d\omega_1$ is the solid angle of the reference surface ΔA_2 subtended at ΔA_1:

$$d\omega_1 = \cos\phi \, dA_2/s^2. \qquad (5\text{-}27)$$

Equation (5-26) is valid only in the limit as $d\omega_1$ approaches zero. This is equivalent to letting s become very large.

As observed from the reference area ΔA_2, the distance s is so great that the relative size of ΔA_1 is negligible and it can be regarded as a point source.

Then the areance [irradiance, illuminance] E at the reference surface is given by definition as

$$E = d\Phi/dA_2 = I\ d\omega_1/dA_2 = I(\cos\phi)/s^2 \quad [\Phi\ m^{-2}]. \quad (5-28)$$

Strictly speaking a real source cannot be treated as a point source obeying the inverse square law unless $s = \infty$. From a practical consideration s must be large enough to yield areance values in agreement with Eq. (5-28) to whatever degree of accuracy is required. Here again the rule of thumb applies that states [8] that for about 1% accuracy, the distance s must be about 10 to 20 times the maximum lateral extension of either the source or the reference surface.

A sample calculation of the solar constant based on the sun as a point source is given as follows: The radiant pointance I of the sun is given by the ratio of the total power to the solid angle (4π sr of a sphere) which is

$$I = P/4\pi = 3.02 \times 10^{25} \quad [W\ sr^{-1}]. \quad (5-29)$$

By Eqs. (5-28), (5-29), and (5-21), the areance [irradiance] is

$$E = I/s^2 = P/4\pi s^2 = L_s\ 4\pi^2 r^2/4\pi s^2 = L_s\ \pi r^2/s^2 = 1340 \quad (5-30)$$

which agrees with the value previously calculated in Eq. (5-22).

Another sample calculation which illustrates the use of a blackbody calibration standard is as follows: A small area blackbody is placed 10 m (on axis so that $\cos\theta = 1$) from the entrance aperture of a spectrometer to obtain a spectral radiant areance [irradiance] calibration at 1.24 μm. The spectrometer records a 4.3-V output signal in response to a 0.0125-in. diameter, 1000K blackbody. The objective of this exercise is to find the instrument responsivity $R(\lambda)$, which is the ratio of output voltage to the spectral areance from the point source.

The solution to Planck's equation for the blackbody source yields $L_b(\lambda) = 3.70 \times 10^2$ which has the units of watts per square meter steradian micrometer at 1.24 μm. By Eqs. (5-26) and (5-28),

$$E(\lambda) = L_b(\lambda)A_s/s^2 = I_b(\lambda)/s^2 = 2.93 \times 10^{-7} \quad [W\ m^{-2}\ \mu m^{-1}]. \quad (5-31)$$

The instrument actually responds to the total power incident upon the detector; however, it can be calibrated in terms of the spectral radiant areance [irradiance] which is related to the spectral power by a constant.

Another interpretation of the distant small area calibration is as follows: Based on Eq. (5-31) the sensor responds to the spectral areance at the reference surface (the instrument aperture).

In either case, the sensor spectral areance responsivity is

$$R_E(\lambda) = V/E(\lambda) = 1.47 \times 10^7 \quad [V(W\ m^{-2}\ \mu m^{-1})]. \quad (5-32)$$

It is significant that the distant small area source spectral areance, at the sensor aperture, is independent of the sensor throughput.

5-8   THE ENTITY OF AREANCE [EXITANCE]

The areance [exitance] M which has the units of watts per square meter may be used to describe a source, and can be derived from the definition of the basic entity sterance [radiance, luminance] as follows: Consider an incremental radiating surface ΔA_s, small compared to a unit hemisphere, that has a sterance L (flux per square meter steradian) as illustrated in Fig. 5-4.

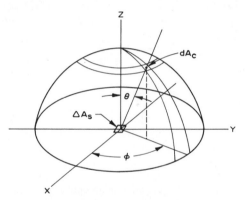

Figure 5-4   Geometry for radiation of an elemental area into a hemisphere.

The elemental collecting surface dA_c on the hemisphere is exactly equal to the elemental solid angle $d\omega$. Therefore, the elemental flux falling on dA_c is

$$d\Phi = L\, d\omega \cos\theta\, \Delta A_s \quad [\Phi]. \qquad (5\text{-}33)$$

Although the factor $\cos\theta$ is associated with the projected area, the projected solid angle is given by $\cos\theta\, d\omega = d\Omega$ so that

$$\Phi = L\, \Delta A_s \int_{(\text{hemisph})} \cos\theta\, d\omega = L\, \Delta A_s \pi \quad [\Phi], \qquad (5\text{-}34)$$

since the projected solid angle Ω for a hemisphere is π. The total flux radiated by ΔA_s is also given by $\Phi = M\, \Delta A_s$ so that

$$M = \pi L \quad [\Phi\ \mathrm{m}^{-2}]. \qquad (5\text{-}35)$$

The constant relating areance M with units of watts per square meter to sterance L with units of flux per square meter steradians is π rather than 2π.

This is a consequence of Lambert's law, or the equivalent fact that L is defined on the basis of projected area.

5-9   THE ENTITY OF STERISENT (EMISSION IN A GAS)

The entities of *sterisent* L^*, of volume emission rate σ, and of the rayleigh are provided to calculate emission, absorption, and scattering in a gas and to calculate the flux incident upon an instrument aperture after propagation through the gas.

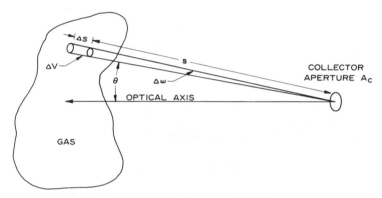

Figure 5-5   Radiation in an optically thin gas.

Consider an incremental volume element ΔV in an optically thin gas (see Fig. 5-5) that has an optical transmissivity $\tau(s)$ (unitless), a transition probability rate A with units of quanta per second, and a number density n with units of inverse cubic meters for a given emission species: The product An is the *photon volume* emission rate σ_p:

$$\sigma_p = An \quad [\text{q sec}^{-1}\ \text{m}^{-3}]. \tag{5-36}$$

If we assume isotropic radiation, the *photon sterisent* L_p^* is given by

$$L_p^* = \sigma_p/4\pi \quad [\text{q sec}^{-1}\ \text{m}^{-3}\ \text{sr}^{-1}]. \tag{5-37}$$

The effective incremental photon rate $\Delta\Phi_p$ incident upon the instrument aperture is given by

$$\Delta\Phi_p = L_p^*\tau(s)\,\Delta V \cos\theta A_c/s^2 \quad [\text{q sec}^{-1}], \tag{5-38}$$

where s is the distance from A_c to ΔV and the transmissivity $\tau(s)$ takes into account losses due to scattering and absorption.

The incremental volume can be expressed in spherical coordinates as

$$\Delta V = s^2 \, \Delta s \, \Delta \omega \quad [\text{m}^3], \tag{5-39}$$

where $\Delta \omega$ is the solid angle of the incremental volume subtended at the collector aperture.

Thus the total flux incident upon the aperture is given by integrating along s and over ω to yield

$$\Phi_p = A_c \int_\omega \left[\int_s L_p * \tau(s) \, ds \right] \cos \theta \, d\omega \quad [\text{q sec}^{-1}], \tag{5-40}$$

where

$$\int_\omega \cos \theta \, d\omega = \Omega_c \quad [\text{sr}]. \tag{5-41}$$

Thus

$$\Phi_p = A_c \Omega_c \int_s L_p * \tau(s) \, ds \quad [\text{q sec}^{-1}]. \tag{5-42}$$

The atmospheric photon sterance L_p is given, according to Eq. (5-19), by the ratio of the flux Φ to the throughput $A_c \Omega_c$, so that

$$L_p = \int_s L_p * \tau(s) \, ds \quad [\text{q s}^{-1} \, \text{m}^{-2} \, \text{sr}^{-1}], \tag{5-43}$$

which gives the photon sterance in terms of the photon sterisent.

A convenient unit for atmospheric studies is the *rayleigh* which is given by combining Eqs. (3-11), (5-37), and (5-43) to obtain

$$L_p' = 4\pi \times 10^{-10} L_p = 10^{-10} \int_s \sigma_p \tau(s) \, ds \quad [\text{R}]. \tag{5-44}$$

The rayleigh may be viewed, in terms of the left-hand side of Eq. (5-44), as the photon sterance (multiplied by a constant) as though the atmosphere were a radiating *surface*. Conversely, the rayleigh may be viewed, based on the right-hand side of Eq. (5-44), as the line integral of the photon emission rate σ_p along the cone of the instrument field of view.

The right-hand side of Eq. (5-44) is especially useful in atmospheric work. A rocket-borne sensor, oriented with the optical axis vertical along the rocket flight path, provides a measure of the rayleigh as a function of s (height) as the rocket passes through the emitting atmosphere. The derivative of the curve of L_p' [Eq. (5-44)] with respect to height s yields the vertical distribution of the photon volume emission rate

$$\sigma_p = An = (10^{10}/\tau_s)(dL_p'/ds) \quad [\text{q sec}^{-1} \, \text{m}^{-3}]. \tag{5-45}$$

The vertical distribution of the number density n (cubic meters) of a particular radiating atmospheric constituent can be inferred from *in situ* measurements of the optical emissions of that constituent, provided the transition probability rate A is known from theory or from laboratory measurements.

5-10   CONFIGURATION FACTORS

There are means to obtain an exact evaluation of Eq. (5-14) for cases where the approximations just given are not adequate. The factors used are known by a variety of names: angle factor, angle ratio, configuration factor, geometrical configuration factor, interchange factor, shape factor, view factor, form factor, or exchange coefficient [9]. Configuration factors are widely used in engineering literature and there are extensive tables for use in evaluating them [10, 11].

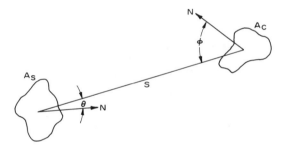

Figure 5-6   Configuration for interchange between two arbitrarily shaped surfaces.

The configuration factor is defined in reference to Fig. 5-6 as the ratio of the flux Φ_{s-c} intercepted by a distant collecting surface A_c to the total flux Φ_t radiated by a source A_s:

$$F_{s-c} = \Phi_{s-c}/\Phi_t = \left(L \int_{A_s} \int_{\omega_s} dA_s \cos\theta \, d\omega_s\right)\Big/\left(L\pi \int_{A_s} dA_s\right) = T/\pi A_s.$$
(5-46)

Since

$$\int_{A_s} \int_{\omega_s} dA_s \cos\theta \, d\omega_s = T.$$
(5-47)

The throughput is therefore given by

$$T = F_{s-c}\pi A_s.$$
(5-48)

The factor F_{s-c} is tabulated for a variety of configurations where either source or receiver is finite or infinitesimal.

REFERENCES

1   Self-Study Manual on Optical Radiation Measurements. Part 1—Concepts. *Natl. Bur. Stand. (U.S.), Tech. Note* No. 910-2, p. 10 (1976).
2   F. E. Nicodemus, Radiometry. *In* "Optical Instruments," Part 1 (R. Kingslake, ed.), Applied Optics and Optical Engineering, Vo.. 4, Ch. 8. Academic Press, New York, 1967.
3   W. H. Steel, Luminosity, throughput, or etendue? Further comments. *Appl. Opt.* **14**, 704 (1974).
4   F. E. Nicodemus, Radiance. *Am. J. Phys.* **31**, 368 (1963).
5   M. Hercher, "Notes on Photometry—Optical and Radiation Detectors," Vol. 7. Inst. Opt., Univ. of Rochester, Rochester, New York (1966).
6   S. Liebes, Jr., Brightness—on the ray invariance of B/n^2. *Am. J. Phys.* **37**, 932–934 (1969).
7   R. C. Jones, Immersed radiation detectors. *Appl. Opt.* **1**, 607 (1962).
8   F. E. Nicodemus, "Reference Book on Radiometric Nomenclature," pp. 23–25. U.S. Nav. Weapons Cent., China Lake, California (draft copy, personal communications) (1974); see also J. W. T. Walsh, "Photometry," 2nd Ed., Constable, London, 1953.
9   Self-Study Manual on Optical Radiation Measurements. Part I—Concepts. *Natl. Bur. Stand. (U.S.), Tech. Note* No. 910-2, pp. 93–100 (1976).
10 R. Siegel and J. R. Howell, "Thermal Radiation Heat Transfer." McGraw-Hill, New York, 1972.
11 E. M. Sparrow and R. D. Cess, "Radiation Heat Transfer." Brooks, Cole, Belmont, California, 1966.

VI

Engineering Calibration

6-1 INTRODUCTION

The calibration of an electrooptical sensor provides for an engineering evaluation of the sensor performance. It is often necessary to ascertain whether or not a sensor is functioning according to design, prior to making a full commitment to obtain a detailed calibration. Proper operation of the sensor can be determined in a relatively simple process referred to here as an "engineering calibration" which might properly be referred to as the final step in the design, fabrication, and testing of a new sensor system.

The evaluation of an electrooptical sensor is based on detector–transducer performance characteristics and system design parameters (or specifications). Therefore, this chapter provides a brief description of detector types, detector parameters, calibration procedures, and sensor performance, but does not provide an exhaustive treatment of these subjects. However, the references provide sources of detailed information should the reader desire more information on these subjects.

6-2   DETECTOR TYPES

Optical sensor designs incorporate two basic detector types: thermal and photon [1]. Thermal detectors respond to heat or power, and therefore have a uniform power response function of wavelength or wave number over the spectral region for which the absorbtance is near unity.

Examples of thermal detectors are the thermopile which is based on the thermoelectric effect, the bolometer which is based on the change in electrical resistance with temperature, the Golay cell which is based on the change in the pressure of a heated gas, and the relatively new pyroelectric detector which is based on the change in polarization of a crystal when it undergoes a variation in temperature [2].

The photon detection process is based on the freeing of charge carriers by the absorption of a single quanta of radiation. They, therefore, respond to the number of effective photons. Photodetectors, consequently, tend to have a linear spectral response function of power versus wavelength.

Examples of photon detectors are the photographic plate which is based on the absorption of a quanta that releases electrons which in turn activates the silver halide in the emulsion, the multiplier phototube which is based on

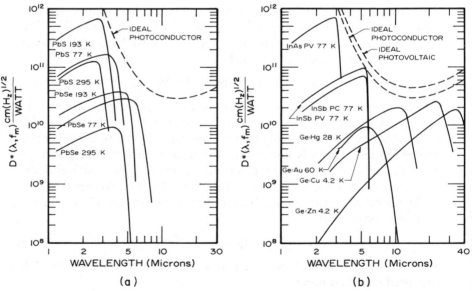

Figure 6-1   Spectral detectivities of ir detectors for various compounds and temperatures: (a) PbS and PbSe, (b) InAs, InSb, and Ge compounds. (Adapted from Santa Barbara Research Center (SBRC) brochure No. 67CM.)

the photoemission effect, the photoconductive detector which is based on the production of free electrons in a solid, of which lead sulfide, lead telluride, lead selenide, germanium, and silicon semiconductors are typical, the photovoltaic which is based on the generation of a photovoltage across a p–n junction in a semiconductor, and the photoelectromagnetic detector which is based on the Hall effect.

The spectral detectivity (defined in the next section) for a variety of photoconductive and photovoltaic detectors is given in Fig. 6-1, and typical photocathode spectral response characteristics for multiplier phototubes are given in Table 6-1 and Fig. 6-2 for purposes of illustration.

6-3   DETECTOR PARAMETERS AND CALIBRATION

From an engineering point of view, the detector parameter of interest is the *noise equivalent power* NEP (or photon rate). Of equal interest is the detector *responsivity*. Generally, the responsivity and the detector noise are measured in two separate tests and the NEP calculated from the results.

The responsivity is defined as the ratio of the output voltage V_s (or current) to the radiant power P_i incident upon the detector. However, both the electrical signal and the incident power are qualified in terms of the rms value of the fundamental component at the chopping frequency [3]. By rms power, in this case, is meant the rms value of the wave form of the chopped radiation. The responsivity is given by

$$R = V_s(\text{rms})/P_i(\text{rms})   [\text{V W}^{-1}]. \tag{6-1}$$

The detector noise generally must be measured in connection with a preamplifier. Detectors are sometimes operated under conditions for which the detector noise is essentially zero and by necessity are preamplifier noise limited. In any case it is important to remove the effect of amplifier gain when the noise is referred to the detector terminals. The electrical noise is also qualified in terms of the rms value at the chopping frequency.

The noise equivalent power NEP is defined as the ratio of the rms noise V_n to the responsivity

$$\text{NEP} = V_n(\text{rms})/R   [\text{W}], \tag{6-2}$$

where the magnitude of NEP depends on the noise bandwidth of the measured electrical noise.

The meaning of NEP is as follows: The noise equivalent power is the incident power that will produce an output voltage $V_s(\text{rms})$ equal to the electrical noise voltage $V_n(\text{rms})$.

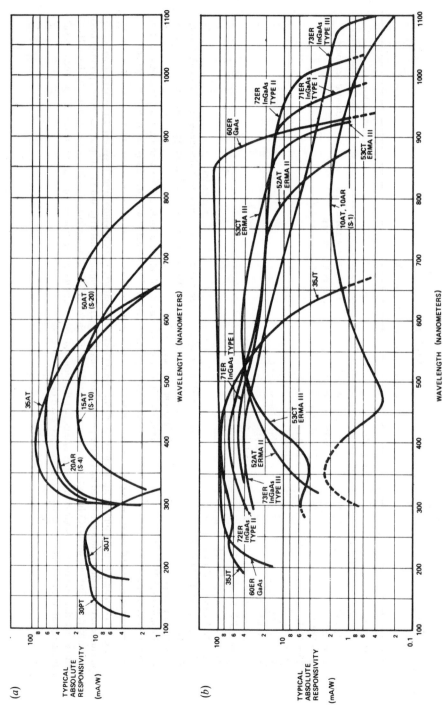

Figure 6-2   Selected spectral response curves for typical multiplier phototube detectors: (a) ultraviolet and visible and (b) broadband. (Adapted from *RCA photomultiplier tubes*, PIT-700C, 12-76.)

TABLE 6-1

Spectral Response Characteristics and Data[a]

Column I	Column II	Column III	Column IV
10 = AgOCs	A = 0080 (lime glass) or	D = dormer-window type	X = extended response
15 = AgBiOCs	7056 (borosilicate glass)	R = reflection type	
20 = CsSb	C = 7740 (Pyrex)	T = transmission type	
25 = CsBi	E = 9741 (uv-transmitting		
30 = CsTe	glass)		
35 = KCsSb (bialkali)	G = 9823 (uv-transmitting		
40 = NaKSb (high	glass)		
temperature bialkali)	J = SiO$_2$ (fused silica)		
50 = NaKCsSb (multialkali)	M = uv-grade sapphire		
51 = NaKCsSb (ERMA I)	P = LiF		
52 = NaKCsSb (ERMA II)			
53 = NaKCsSb (ERMA III)			
60 = GaAs			
71 = InGaAs (type I)			
72 = InGaAs (type II)			
73 = InGaAs (type III)			

Examples:

Tube type 931A has a spectral response that was previously designated as 102 (S-4). This tube type has a CsSb photocathode, a 0080 window, and a reflection type photocathode. Its new designation is 20AR.

Similarly, a tube type having a CsSb photocathode, a 0080 window, and a transmission type photocathode is designated 20AT. This response was previously designated 107 (S-11).

[a] RCA has changed its pure-numeric notation for specifying spectral-response characteristics to a more orderly system. The new designations are alphanumeric combinations that are based on (1) the photocathode material, (2) the window material, and (3) the photocathode operating mode. As illustrated, the first two digits (Column I) in the designation indicate the photocathode material; the following alpha character (Column II), the window material; and the next alpha character (Column III), the photocathode operating mode. Where required, the letter "X" is used as a suffix to the designation to indicate an extended response in the red or near infrared.

There is a large class of detectors for which the noise voltage is proportional to the square root of the detector area A_d. In this special case, the detectivity D^* (pronounced D-star) is a better figure of merit than the NEP. D-star is defined as [4-6]

$$D^* = (A_d\, \Delta f)^{1/2}/\text{NEP}   [\text{m Hz}^{1/2}\ \text{W}^{-1}]. \tag{6-3}$$

The parameter D^* can be considered as the inverse noise equivalent power that has been normalized for detector area and electrical noise bandwidth. The spectral D-star $D^*(\lambda)$ is the detectivity at a particular wavelength λ. The responsivity of photon detectors tends to be a linear function of wavelength (see Fig. 6-1). The value of $D^*(\lambda)$ will have its maximum value near the long wavelength cutoff λ_p.

D-star can also be defined in terms of the responsivity by combining Eqs. (6-2) and (6-3) as

$$D^* = (A_d\, \Delta f)^{1/2} R/V_n. \tag{6-4}$$

Several methods are used to measure $D^*(\lambda)$. One method is to allow a measured amount of radiation in a narrow spectral interval to fall upon the detector. However, it is difficult to measure the absolute intensity of this monochromatic radiation. A more commonly used method is to measure the blackbody $D^*(T)$, usually with a 500K source, and then calculate $D^*(\lambda)$ as [6]

$$D^*(\lambda) = kD^*(T). \tag{6-5}$$

The blackbody responsivity is given by

$$R(T) = V_s(\text{rms})\Big/\int_0^{\infty} P_i(\lambda)\, d\lambda(\text{rms})   [\text{V W}^{-1}]. \tag{6-6}$$

The value of the integral in the denominator is the incident power at all wavelengths including those regions beyond λ_p where it is not effective in producing an output in the detector; however, it can be calculated directly from the Stefan–Boltzmann law and the associated geometry, and does not need to be measured.

The detector responsivity may be defined as

$$R = R(\lambda_p)R(\lambda), \tag{6-7}$$

where $R(\lambda_p)$ is the peak responsivity in units of volts per watt and $R(\lambda)$ the relative (peak normalized) responsivity (unitless), and $R(\lambda_p) > R(T)$. The detector output $V_s(\text{rms})$ in response to the incident blackbody power is given by

$$V_s(\text{rms}) = R(\lambda_p) \int_0^{\infty} R(\lambda)P_i(\lambda)\, d\lambda   [\text{V}]. \tag{6-8}$$

Thus a knowledge of the detector relative spectral responsivity $R(\lambda)$ is required, but is relatively easy to measure.

The ratio of $R(\lambda_p)$, from Eq. (6-8), to $R(T)$, from Eq. (6-6), yields a correction factor k:

$$R(\lambda_p)/R(T) = D^*(\lambda_p)/D^*(T) = \int_0^\infty P_i(\lambda)\, d\lambda \bigg/ \int_0^\infty R(\lambda)P_i(\lambda)\, d\lambda = k \quad (6\text{-}9)$$

which can be used to find $D^*(\lambda_p)$ since

$$D^*(\lambda_p) = kD^*(T). \tag{6-10}$$

The detector responsivity $R(\lambda)$ tends to be a linear function of wavelength for photon detectors up to λ_p where it falls rapidly to zero (see Fig. 6-1). Thus the normalized responsivity can be approximated by

$$R(\lambda) \simeq \begin{cases} \lambda/\lambda_p & \text{for} \quad \lambda \leq \lambda_p \\ 0 & \text{for} \quad \lambda > \lambda_p \end{cases} \tag{6-11}$$

Then Eq. (6-9) becomes

$$k \simeq \int_0^\infty \lambda^{-5}[\exp(hc/\lambda kT) - 1]^{-1} \bigg/ \int_0^\infty (\lambda^4\lambda_p)^{-1}[\exp(hc/\lambda kT) - 1]^{-1},$$
$$\tag{6-12}$$

where Planck's equation is used to calculate the incident power.

Often $R(\lambda)$ is not analytic, and computer methods are used to calculate k, where the integrals are approximated as

$$k \simeq \sum_{i=1}^n P_i(\lambda) \bigg/ \sum_{i=1}^n R_i(\lambda)P_i(\lambda) \tag{6-13}$$

with midpoint values used for each increment.

The solution of Eq. (6-12) for photon detectors is given in graphical form in Fig. 6-3 which provides values of k that are in common use by manufacturers of IR detectors.

6-4   NOISE EQUIVALENT POWER

The "sensitivity" of an optical instrument is best expressed as a ratio of signal to internal noise rather than simply as a ratio of input to output. Accordingly, the power sensitivity is specified as the *noise equivalent power* (NEP). The term "noise equivalent" is interpreted as an input power that produces an output signal (a change in the mean output) equal to the standard deviation of the output noise.

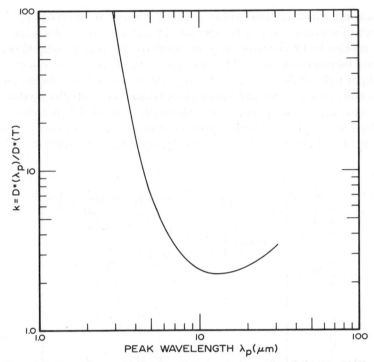

Figure 6-3 Dependence of factor k on detector peak wavelength for $T = 500K$.

The sensitivity of a radiometer or photometer to a point source is commonly specified as the *noise equivalent flux density* (NEFD), and the sensitivity of an instrument to an extended source is the less common *noise equivalent sterance* [radiance] NES [NER]. However, the standard deviation is dependent on the "integration time"; therefore, it is not meaningful to compare instruments directly.

The spectrometer scans over the *free spectral range* to give a continuous function which represents the wavelength of a source. This function is referred to as the *power spectral density function* or, simply, the *spectrum.* The sensitivity of a spectrometer is characterized in terms of *noise equivalent spectral sterance* [radiance] NESS [NESR] for extended sources, and the less common term *noise equivalent spectral areance* [irradiance] NESA [NESI] for point sources.

The degree of difficulty of the calibration of an optical instrument is a function of its sensitivity. This is especially true of instruments that utilize multiplier phototube detectors or of cryogenically cooled IR detectors.

A multiplier phototube can be damaged if it is exposed to ambient levels of light while energized. An instrument utilizing a multiplier phototube generally must be calibrated within a special darkroom that is equipped with calibration sources that can be operated at suitably controlled levels.

A cryogenic IR detector may be sensitive to visible light as well as ambient thermal emissions. The detector must therefore be shielded in a cold, light-tight enclosure. Special cryogenically cooled chambers that provide a cold background and light-tight environment must be used to calibrate instruments that utilize cryogenically cooled IR detectors. The chambers must provide both cryogenic and vacuum continuity with the instrument and are expensive to fabricate, operate, and maintain.

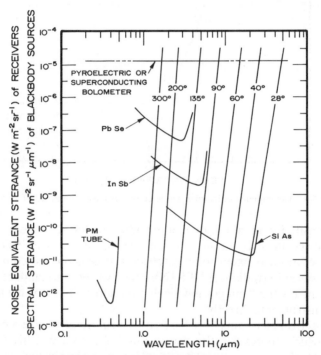

Figure 6-4   Noise equivalent sterance [radiance] for typical state-of-the-art sensors that have been normalized to unity relative aperture, noise bandwidth, optical transmittance, and chopping factor; also blackbody sterance [radiance] as a function of wavelength and temperature.

Figure 6-4 gives the noise equivalent sterance [radiance] NES [NER]—the sensitivity to an extended area source—for typical state-of-the-art instruments that have been normalized to unity throughput, unity electrical bandwidth, and unity optical transmittance and chopping factor. Also plotted in Fig. 6-4 is a family of blackbody curves that gives the sterance [radiance] as a function of wavelength and temperature. Figure 6-4 is interpreted as follows: most state-of-the-art IR instruments will exhibit excessive response to 300K ambient background beyond 1 μm. Figure 6-4 indicates that state-of-the-art cryogenic instruments are capable of detecting energy emanating from blackbody-type sources at less than 30K provided the optical throughput and integration time are sufficiently large. Multiplier phototubes exceed all other detector types in sensitivity, but are limited to the visible or near visible region of the spectrum.

The instantaneous value of the noise frequently exceeds the standard deviation; thus a realistic "minimum detectable signal" is roughly three to five times the NEFD or NES since the deflection must be measured against the instantaneous random noise.

6-5   INSTRUMENT SENSITIVITY

The major objectives of the calibration of an electrooptical instrument relate to the measurement of incident flux from a remote source of flux. However, from an engineering point of view, another application of calibration is associated with the evaluation of the performance of an electrooptical instrument. The engineering calibration might properly be considered the final test phase of performance evaluation.

Often, the sensitivity of the instrument is the parameter of interest. State-of-the-art design requires the suppression of excess noise forms, and instrument sensitivity is one of the parameters indicative of good design and careful fabrication.

The sensitivity is dependent on parameters such as the optical throughput; the transmittance and reflectance of mirrors, lenses, filters, gratings, etc.; the electrical or noise bandwidth (or time constant); the chopping factor; and the detector sensitivity. These parameters are usually determined by design considerations and are known; therefore, it is possible to predict the system performance.

The sensor noise equivalent sterance [radiance, luminance] sensitivity NES [NER] is obtained by the invariance of sterance by dividing the detector NEP by the sensor throughput T, which is the product of the collector area A and the projected solid angle field of view Ω.

It is also necessary to take into account the optical losses introduced by the chopper (when used) and by the reflection and transmission losses of mirrors, lenses, gratings, etc. Thus

$$\text{NES} = \beta\text{NEP}/T\tau \quad [\text{W m}^{-2} \text{ sr}^{-1}], \quad (6\text{-}14)$$

where β is the chopping factor and τ the system optical transmittance.

The chopping factor [7] of a square wave chopper (where the amplifier system preserves the square wave form) is 2.0. The chopping factor that results from the common use of a tuned amplifier (where the amplifier passes only the sine wave fundamental component) is 2.46. The instrument which produces a triangular wave resulting from the use of a chopper blade and aperture of the same size has a chopping factor of 4. A dc-reset system operates with a chopping factor that approaches unity [8].

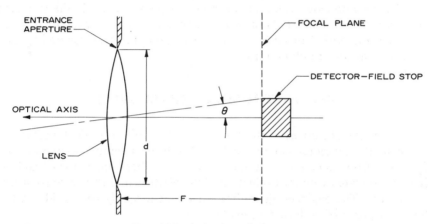

Figure 6-5   A simple optical sensor.

For a simple optical system, in which the detector forms the field stop (see Fig. 6-5), the throughput can be expressed in terms of the system f-number [9] (the ratio of the collector lens effective focal length F to its diameter d) which is represented by the symbol (f-no). The collector area for a circular field of view is given by

$$A = \tfrac{1}{4}\pi d^2 \quad [\text{m}^2] \tag{6-15}$$

and the projected field of view is given approximately by

$$\Omega = A_d/F^2, \tag{6-16}$$

where F is the effective focal length and where A_d the detector (field stop) area. Thus

$$\text{NES} = (4\beta \text{NEP}/\pi A_d \tau)(f\text{-no})^2 \quad [\text{W m}^{-2} \text{ sr}^{-1}]. \tag{6-17}$$

Equation (6-17) is significant in evaluating a system design because the field of view and the system (f-no) are independent parameters. The (f-no) is a very important parameter because it appears to the second power and has a major impact on the instrument sensitivity.

The NEFD of an instrument may be obtained in a similar manner, by dividing the NEP by the collector area A_c:

$$\text{NEFD} = \beta \text{NEP}/A_c \tau \quad [\text{W m}^{-2}]. \tag{6-18}$$

Equations (6-14) and (6-18) can be expressed in terms of the detector D^* rather than NEP. Using Eq. (6-3) the noise equivalent sterance [radiance] is given by

$$\text{NES} = \beta(A_d \, \Delta f)^{1/2}/D^* \tau T \quad [\text{W m}^{-2} \text{ sr}^{-1}] \tag{6-19}$$

and the noise equivalent flux density is given by

$$\text{NEFD} = \beta(A_d \, \Delta f)^{1/2}/D^* A_c \tau \quad [\text{W m}^{-2}]. \tag{6-20}$$

The spectral quantities (NESS and NESA) are obtained by normalizing with respect to the spectral bandpass $\lambda_2 - \lambda_1$.

Considerable work should be done before the engineering calibration begins. Equations are solved to predict, as appropriate, NES or NEFD based on a specified detector NEP or D^*, or conversely the detector NEP or D^* may be calculated from the noise equivalent flux measured in the engineering calibration.

The detailed procedures undertaken in an engineering calibration are similar to those of a final calibration except the objective of the calibration is different. The redundancy necessary to estimate the precision and accuracy in a detailed calibration is not necessary to evaluate instrument performance. For example, the field of view may be approximated by simply measuring the half-power angles and calculating the solid angle field of view based on certain assumptions of symmetry. The absolute responsivity can be measured based on idealized spectral characteristics and a small area distant source. Thus only enough data are obtained to assure proper operation of the sensor.

REFERENCES

1   R. A. Smith, F. E. Jones, and R. P. Chasmar, "The Detection and Measurement of Infrared Radiation." Oxford Univ. Press, London and New York, 1957.

2   H. Blackburn, and H. C. Wright, Thermal analysis of pyroelectric detectors. *Infrared Phys.* **10**, 191 (1970).

3   R. C. Jones, D. Goodwin, and G. Pullan, "Standard Procedure for Testing Infrared Detectors and for Describing Their Performance." Off. Dir. Def. Res. Eng., Washington, D.C., 1960.

4   R. C. Jones, Detectivity: The reciprocal of noise equivalent input of radiation. *Nature* (*London*) **170**, 937–938 (1952).

5 R. C. Jones, Performance of detectors for visible and infrared radiation. *Adv. Electron.* **5**, pp. 1–96 (1953).

6 P. W. Kruse, L. D. McGlaughlin, and R. B. McQuistan, "Elements of Infrared Technology," pp. 362–363. Wiley, New York, 1962.

7 C. L. Wyatt, An optimum system synthesis for optical radiometric measurements. Ph.D. Thesis, Utah State Univ., Logan, 1968.

8 C. L. Wyatt, Infrared spectrometer: Liquid-helium-cooled rocketborne circular-variable filter. *Appl. Opt.* **14**, 3086 (1975).

9 F. A. Jenkins and H. E. White, "Fundamentals of Optics," pp. 114, 170. McGraw-Hill, New York, 1957.

VII

Standards and Calibration Uncertainty

7-1   INTRODUCTION

Throughout this book it is assumed that standard radiant sources are available, and that the appropriate field entities can be calculated by the geometry of radiation transfer. The objective of the calibration of an electro-optical sensor, that the measurement be independent of the sensor, requires that absolute or traceable standards be available for use by the different observers who make such measurements from time to time. The calibration report should always include a statement of the estimated uncertainty in the calibration.

The objective of this chapter is to briefly describe radiometric standards and the role of the National Bureau of Standards (NBS) in providing measurement services in the United States. The problems of ascertaining the precision (repeatability) and the accuracy of the absolute calibration of electrooptical sensors are also considered.

7-2 THE NATIONAL BUREAU OF STANDARDS

NBS' stated goal in radiometry and photometry is to make 1% measurements commonplace in the United States [1]. There has therefore been an ongoing program to develop standards and calibration techniques directed toward the achievement of this goal. The Optical Radiation Section of NBS includes a Photometric and Radiometric Dissemination Group with activities that fall into three categories: calibrations, intercomparisons, and developmental projects [2].

7-2-1 Calibrations

NBS provides a variety of calibration procedures which include but are not limited to the following [3, 4]:

1.   Blackbodies may be submitted for calibration at fixed temperatures (500–700K) at a number of wavelengths between 1.5 and 14 μm, with uncertainties between 2% at 1.5 μm and 1% at 14 μm as part of their "basic" and "gage" calibration service [5]. Special calibrations having unique requirements, not satisfied by the listed basic and gage items, are considered as small research efforts. Details of the calibration procedure and a summary of the documentation of the uncertainties are provided in each calibration report.

2.   Optical pyrometers may also be submitted for calibration.

3.   Standard lamps are used for the visible and ultraviolet because open blackbodies cannot be operated at sufficiently high temperatures. Lamps may be submitted to NBS for calibration or calibrated lamps may be supplied by NBS as follows:

a.   Ribbon filament lamps are calibrated at numerous wavelengths from 0.255 to 2.4 μm. These lamps are qualified in terms of spectral radiant sterance [radiance] with uncertainties that vary from 1 to $4\frac{1}{2}$%. These lamps require 40 A dc at 12 V.

b.   Quartz–halogen lamps are available as spectral areance [irradiance] standards, calibrated at numerous wavelengths from 0.25 to 1.6 μm, with uncertainties that vary from 1.5 to 5%.

c.   Tungsten filament lamps, with inside frosted bulbs, are available for luminous pointance [intensity] standards. These bulbs operate at a color temperature of about 2700K with an uncertainty of about 4%.

d.   Gas-filled lamps are available as luminous flux (geometrically total) standards. They have an approximate color temperature of 2500 to 3000K, with uncertainties of 1.5 to 4.5%.

e.   Deuterium arc lamps are available for use as ultraviolet standards.

7-2-2   Intercomparisons

Part of the NBS program is to assess and verify measurements made in laboratories other than NBS. This is accomplished by using various transfer standard sources and detectors. Calibrated detectors can be rented from NBS for use as transfer standards [6]. An electrically calibrated pyroelectric radiometer (ECPR) was recently compared with three other absolute radiant power measurement systems, two in Europe [the Physikalisch–Technische Bundersanstalt (PTB) in West Germany and the World Radiation Center (WCR) in Davos, Switzerland], and one at NBS (Boulder, Colorado). The European detectors were both cavity-type, electrically calibrated, thermopile radiometers and the NBS detector was a calorimeter type. Differences among measurements made with the ECPR and each of the other three systems were 0.6% and less [6, p. 1].

7-2-3   Developmental Projects

NBS has a program to improve the quality and extend the range of calibration services available. A new calibration facility for calibrating blackbody sources for temperatures from -40 to $+100$ °C has been completed [7]. The capability for calibrating the sterance [radiance] of small, low-temperature blackbodies in a vacuum at 20K is being developed [8].

The calibration of low-temperature blackbodies is based on the calorimeter approach wherein the temperature rise in a cavity receiver is compared with that produced by an electrical heater located within the cavity. This method requires that the receiver and the source be operated in a high-vacuum cryogenic environment. The receiver is thermally coupled to an infinite (cold) heat sink by a finite thermal resistance so that a steady radiant power input results in an equilibrium temperature rise. The radiation is periodically blocked with a shutter and then an internal heater is activated. The electrical power is adjusted until it produces a temperature equal to that resulting from the radiant power. The radiant power is equal to the electrical power measurement if the heat loss due to the heater and temperature probe leads are taken into account.

The Optical Radiation Section of NBS is working on a *Self-Study Manual on Optical Radiation Measurements* to aid technical personnel in a variety of disciplines to perform accurate optical radiation measurements. This work, directed at the bachelor of science level of understanding, is to include concepts, instrumentation, and applications in such a way as to contribute to the NBS goal of 1% accuracy in optical radiation measurements [9].

7-3   STANDARDS

Accuracy implies absolute standards and traceable standards. The concept of traceable standards, or secondary standards, has less meaning for radiation standards than it does for the standard meter, for example. The international standard of length is a bar of a platinum–iridium alloy called the *standard meter*, which is kept at the International Bureau of Weights and Measures near Paris, France. Because of the inconvenience of this location, copies of this primary standard are made, called *secondary standards*, which are distributed to various standardizing agencies such as the National Bureau of Standards in Washington, D.C. Thus a measure of length can be traced through secondary standards to the primary standard.

The inconvenience associated with the use of primary and secondary standards has led to the search for a more fundamental definition of the standard length based on a physical process that can, at least in principle, be reproduced in any laboratory.

The distance between the two lines on the standard bar have been precisely compared with the wavelength of orange light emitted by atoms of a single pure isotope, krypton-86, in an electrical discharge. Krypton-86 atoms are universally available and the orange light from them can be reproduced in any laboratory. The wavelength of this light is therefore a truly accessible standard. The meter is defined as 1.65076373×10^6 times this wavelength.

A primary radiation standard comparable with the standard meter does not exist. Radiant sterance [radiance], which has the units of watts per square meter steradian, is a derived rather than a fundamental entity. However, it does depend on the fundamental unit of length. The performance of blackbody radiators can be predicted based on fundamental physical processes. Thus the blackbody has been employed as a standard source of radiation flux, and is a source which can in principle be reproduced in any laboratory and can be traced to standards of length and absolute temperature.

Blackbody sources are comparatively easy to fabricate and employ as standards for microwaves, infrared, and long-wavelength visible radiation. However, they are difficult to use at shorter wavelengths because of the small amount of short-wavelength radiation produced at practical temperatures. Therefore, high-temperature greybody sources, such as tungsten lamps, are used for visible and ultraviolet standards. The greybody source can be calibrated by comparison with a blackbody source in the regions of overlapping wavelength, or through the use of a standard detector.

Standard detectors are also used as absolute and transfer standards. An electrically calibrated detector of radiant energy can be traced to standards of electrical voltage, current, and power.

7-3-1 Standard Sources

The spectral radiant sterance [radiance] of a blackbody is given by Planck's equation, which is dependent on the wavelength and the absolute temperature. The *effective flux* arriving at the instrument aperture depends on the sensor spectral response and the blackbody temperature. It may also be dependent on certain geometrical properties of a particular physical calibration experiment. The standard source parameters must be varied in order to fully characterize the sensor; this means that the source uncertainty must be qualified under an infinite number of operating conditions.

Specifications of commercial blackbody sources generally give the uncertainty of the temperature and the emissivity as well as the control settability and stability. Typical units can be set to an accuracy of $\pm 0.25°C$, have a long-term stability of $\pm 0.5°C$, and an emissivity of $99.0 \pm 1.0\%$.

The uncertainty of the spectral radiant sterance [radiance] of a blackbody source can be determined by solving Planck's equation over the uncertainty of the temperature. For example, a 500K blackbody, which peaks at 5.8 μm, has an uncertainty of 5.9% at 1 μm and only 0.6% at 10 μm over a temperature uncertainty of 1K.

A "certificate of calibration" is usually provided for the temperature-measuring thermocouple of commercial blackbodies that gives the uncertainties in the listed value. It is often possible to remove the thermocouple so that it can be sent to NBS to be certified as to its absolute accuracy. However, the temperature of the blackbody may not correspond exactly to the indicated temperature of the thermocouple.

The optical pyrometer is a device that permits visual measurement of the blackbody source temperature. Accuracies are possible to $\pm 0.1K$ [10]. However, the optical pyrometer cannot be used to measure the temperature of blackbody sources that are not hot enough to emit visible radiation.

7-3-2 Standard Detectors

Another approach to primary radiometric standards is obtained through the use of a well-qualified receiver or detector of radiant energy. Standard thermal detectors that are absolute or self-calibrating are available. They are blackened to obtain an absorptance as close to unity as possible. This is sometimes accomplished by employing some form of cavity configuration with well-blackened opaque walls [11]. Provision is made to heat the receiver electrically to produce the same heating effect as does the incident radiated power. Thus the measurement of spectral power is based directly on electrical standards without recourse to separate primary radiometric standards. It has long been established that electronic measurements of voltage, current, and power can be made at very low levels of uncertainty. The technique of the electrically calibrated receiver can, in principle, be applied to any thermal detector such as the thermocouple, thermopile, or calorimeter.

Until recently this technique has been limited by the rather poor sensitivity of conventional thermal detectors; however, the development of a null-balance, electrically calibrated, pyroelectric detector has to some extent overcome this limitation. The emergence of the electrically calibrated detector as a practical alternative to standard sources for use in a wide range of optical calibrations offers the possibility for a system of standardization that has great flexibility and provides for a short chain of traceability for many radiometric measurements [12].

7-4   STANDARDS OF WAVELENGTH

Standards of wavelength for the calibration of monochromators and spectrometers can be reproduced in any laboratory based on either absorption or emission of various gases. The discharge line-spectrum of mercury, for example, is available in the emission of a common commercial fluorescent lamp which utilizes mercury vapor. These lines form excellent wavelength standards in the visible range. Absorption cells filled with ammonia vapor or carbon dioxide serve well for the IR range. The *Handbook of Chemistry and Physics* lists a large number of wavelengths of known spectral lines with more than adequate accuracy for many applications. Certain plastics exhibit characteristic absorption bands due to chemical bonds, for example, polystyrene, which can be identified and used for wavelength standards in the IR.

7-5   CALIBRATION UNCERTAINTY

The calibration of an electrooptical sensor can be regarded as an experiment in which the output is obtained in response to a standard source. It is observed that a series of such experiments will yield different values. An observation of the ensemble of measurements indicates that the output can be regarded as a random variable x.

The *mean* of such a variable is given by [13]

$$\bar{x} = \int_{-\infty}^{\infty} xP(x)\, dx \tag{7-1}$$

for a continuous function and

$$\bar{x}_s = (1/n) \sum_{i=1}^{n} x_i, \tag{7-2}$$

where the probability density function $P(x) = 1/n$ (for equally probable measurements) for the *sample* mean. The sample mean is an estimate of the "true" mean $\bar{x}$ [13, pp. 18–19].

The spread of the measured values of x_i about $\bar{x}$ is characterized by the variance σ^2 of x:

$$\sigma^2 = \int_{-\infty}^{\infty} x^2 P(x)\, dx - \bar{x}^2 \tag{7-3}$$

for continuous functions [13] and

$$\sigma_s^2 = 1/(n-1) \left[\sum_{i=1}^{n} x_i^2 - n\bar{x}_s^2 \right] \tag{7-4}$$

where $n - 1$ is used rather than n for the *sample* variance. The subscript s, denoting sample mean, will not be used hereafter. The sample variance is an estimate of the "true" variance.

The probability density function of the variable x may be unknown, and there may be a finite number n of observations x_i obtained in the calibration exercise. However, according to the law of large numbers [14], the mean and variance become better estimators of the parent random process as the number of measurements is increased without limit.

The central limit theorem [14, p. 81] gives assurance that the probability distribution function of the sample mean becomes Gaussian as the number of statistically independent samples is increased without limit, regardless of the probability distribution of the parent random process, provided it has a finite mean and variance.

The error, at any instant of time, can be described in terms of a probabilistic statement based on the standard deviation σ. The data in Table 7-1, computed from the normal function [15], indicate that a 95% confidence level is obtained for 2σ. This means that the variable will remain within $x \pm 2\sigma$ some 95% of the time.

TABLE 7-1

The Probability That the Error
Will Not Exceed $\pm\sigma$

Uncertainty	Probability (%)
σ	68.26
1.5σ	86.64
2.0σ	95.44
2.5σ	98.76
3.0σ	99.74

7-5-1 Precision and Accuracy

One of the major goals of the calibration of an electrooptical sensor is to determine, as far as possible, the extent of the errors associated with any measurement. The error, defined here as the difference between the indicated and "true" value of the entity measured, results from various random and systematic processes.

The effect of random errors can often be reduced by repeated measurement or by averaging, which increases the *precision* of a measurement. However, the effect of systematic errors cannot be reduced by averaging. Thus a measurement may be precise but at the same time inaccurate.

The difference between the *mean* indicated value and the *true* value of an entity measured is a measure of the accuracy. Thus a measurement may be accurate although imprecise. This is illustrated as follows: Even in the presence of large random errors an accurate measurement may be repeated sufficiently many times to give the true value. This is not the case with systematic errors, even if the random error is zero. In general, an accurate measurement gives truth; a precise measurement is repeatable.

The *precision* of a measurement is a measure of the reproducibility or consistency of measurements made with the same sensor. The *accuracy* of a measurement refers to absolute measurements and implies absolute standards or traceable standards [13, p. 3; 16]. Neither the precision nor the accuracy can be determined exactly but must be estimated from the observed data and in consideration of the physical properties of the standard sources used. However, such estimates should be part of any statement of measured values [16].

It is often difficult to obtain electrooptical measurements within an uncertainty of 5%. The uncertainty of measurements made with high-vacuum cryogenic systems is probably greater than 50%. These uncertainties are limited by the accuracy with which standard sources and sensor spectral responsivity can be characterized.

7-5-2   Sources of Error

The nature of the application of an electrooptical sensor determines the need for precision and accuracy. For example, target characterization based on ratios of appropriate radiant entities requires only that the measurements be precise, while absolute flux measurements require both precision and accuracy.

Systematic errors must be assumed zero in a standard source since every effort is made to eliminate them; however, the source output is only known to some degree of accuracy. The uncertainty of the source can usually be traced back to the uncertainty of a standard detector, or to absolute temperature and emissivity measurements. The uncertainty of the standard source is characterized by σ_S which includes such factors as resettability, repeatability, and stability.

The radiant input to the sensor is also dependent on the geometrical relationship between the standard source and the sensor. The problem of determining the uncertainty of geometrical variables can sometimes be avoided when, for example, an extended area source is used for an extended area calibration. This again illustrates the value of the calibration rule of good performance that dictates that the calibration of the sensor should be conducted under conditions that approximate the measurement condition.

Assuming that the source is noise-free and exhibits short-term stability, the sensor precision can be considered independently of the source. Then there are two distinct sources of uncertainty associated with sensor precision.

The first source of sensor uncertainty arises from sensor-generated system noise. The standard deviation of the system noise σ_N can be evaluated in an experiment referred to hereafter as the *dark-noise analysis*. The standard deviation of the system noise can be specified as the *change* in the input flux $\Delta\Phi$ that causes the output mean to change by an amount equal to σ_N. This is also the sensor *noise equivalent flux*, and is significant primarily for measurements where the flux levels are near the threshold of system sensitivity.

The second source of sensor uncertainty is determined from experiments referred to hereafter as the *linearity analysis* and the *absolute calibration*. In each case, the quality of fit of the experimental data set to the calibration equation is calculated as a relative error and is characterized as σ_L and σ_W, respectively.

The precision of the sensor, as observed in the calibration is not completely independent of the standard source. The observed spread in the sensor sample data may be dependent on how well the source and sensor spectral characteristics are known. The experiments as outlined hereafter are designed to produce the greatest degree of independence possible. Then the precision of the sensor is given by [13,14]

$$\sigma_P = (\sigma_N{}^2 + \sigma_L{}^2 + \sigma_W{}^2)^{1/2}. \tag{7-5}$$

The accuracy of the sensor calibration is given by

$$\sigma_A = (\sigma_N{}^2 + \sigma_L{}^2 + \sigma_W{}^2 + \sigma_S{}^2)^{1/2}.$$

The succeeding chapters outline the methods used to characterize the sensor response in each domain in a way that tends to result in the desired independence. In each case the error spread is also calculated as the quality-of-fit of the sample data set to the calibration equation as a percent standard deviation.

REFERENCES

1 National Bureau of Standards, U.S. Department of Commerce *Opt. Radiat. News* No. 3, p. 1 (1974). [This newsletter is prepared bi-monthly by the staff of the Optical Radiation Section of NBS to report on items of interest in optical radiation measurements. Inquiries may be directed to A. T. Hattenburg, A223 Physics Bldg., NBS, Washington, D.C. 20234 (301-921-2008).]

2 National Bureau of Standards, U.S. Department of Commerce, *Opt. Radiat. News* No. 2, p. 1 (1974).

3 National Bureau of Standards, U.S. Department of Commerce, *Opt. Radiat. News* No. 9, pp. 1–10 (1975).

4 Calibration and Test Services of the National Bureau of Standards. *Natl. Bur. Stand. (U.S.)*, *Spec. Publ.* No. 250, pp. 14–15 (1976). (Appendix Fees for Services.)

5 Listings from NBS Measurements Users Bulletin No. 4, *Natl. Bur. Stand. (U.S.)*, *Spec. Publ.* No. 250, Suppl. (1970).

6 National Bureau of Standards, U.S. Department of Commerce, *Opt. Radiat. News* No. 17, p. 2 (1976).

7 National Bureau of Standards, U.S. Department of Commerce, *Opt. Radiation News* No. 15, p. 1 (1976).

8 National Bureau of Standards, U.S. Department of Commerce, *Opt. Radiat. News* No. 9, p. 7 (1975).

9 National Bureau of Standards, U.S. Department of Commerce, *Opt. Radiat. News* No. 7, p. 3 (1975).

10 Review of the present state of primary radiation pyrometry. *Rev. Int. Hautes Temp. Refract.* **12**, 172–179 (1975). (Abstr. *Eng. Index* **75**, 5690.)

11 F. E. Nicodemus, Radiometry. *In* "Optical Instruments," Part 1 (R. Kingslake, ed.), Applied Optics and Optical Engineering, Vol. 4, Ch. 8. Academic Press, New York, 1967.

12 Staff Report, Detectors and radiometry. *Opt. Spectra* January, pp. 34–35 (1976).

13 P. R. Bevington, "Data Reduction and Error Analysis for the Physical Sciences," p. 16. McGraw-Hill, New York, 1969.

14 W. B. Davenport, Jr. and W. L. Root, "An Introduction to the Theory of Random Signals and Noise," p. 79. McGraw-Hill, New York, 1968.

15 M. R. Spiegel, "Mathematical Handbook of Formulas and Tables," p. 257. McGraw-Hill, New York, 1968. (The Schaum's Outline Series.)

16 F. E. Nicodemus and G. J. Zissis, "Report of BAMIRAC—Methods of Radiometric Calibration," ARPA Contract No. SD-91, Rep. No. 4613-20-R (DDC No. AD-289, 375), p. 2. Univ. of Michigan, Infrared Lab., Ann Arbor, Michigan (1969).

VIII

Dark-Noise Analysis

8-1 INTRODUCTION

The first step in the calibration of an electrooptical sensor is to qualitatively and quantitatively evaluate the dark-noise signal. Such an evaluation can be obtained when the sensor aperture is blocked with a light-tight cover. The noise content of the dark signal provides for one measure of the sensor precision, or reproducibility, and for a measure of the false signal content of the output.

In general, the properties of the dark-noise output are a function of wavelength, or spectral scan position, of a spectrometer. A radiometer is a degenerate case of the spectrometer for which there is only one wavelength.

The precision, or reproducibility, of the sensor is limited by the system noise. The variance of the dark-noise signal is a measure of the uncertainty in terms of either the output signal or the input flux.

The noise characteristics of a sensor may change with time or environmental conditions. An evaluation of the noise characteristics in connection with a calibration experiment is essential but not sufficient. Any field measurement program should also include an aperture-covered run to provide for a noise analysis under field conditions. For this reason, the calibration equation, described in the next chapter, does not include a constant for the zero intercept (offset).

8-2 THE DARK-NOISE MEAN AND VARIANCE

The statistical characteristics of the dark-noise output are approximated by the sample mean and the sample variance based on a set of measurements.

The mean value and the standard deviation of the dark signal of a spectrometer are obtained by processing many scans over a relatively long time. This may be accomplished by constructing a matrix of the data in digital form where there are n scans, each containing m sample points:

Wavelength	λ_1	λ_2	λ_3	$\cdots$	λ_j	
Scan 1	$x_{1,1}$	$x_{1,2}$	$x_{1,3}$	$\cdots$	$x_{1,j}$	
Scan 2	$x_{2,1}$	$x_{2,2}$	$x_{2,3}$	$\cdots$	$x_{2,j}$	(8-1)
$\vdots$	$\vdots$	$\vdots$	$\vdots$		$\vdots$	
Scan n	$x_{i,1}$	$x_{i,2}$	$x_{i,3}$	$\cdots$	$x_{n,m}$	

The jth column refers to the jth wavelength and the ith row refers to the ith scan. In Eq. (8-1), $x_{i,j}$ may be either the output (voltage or current) or the inverse transform of the output, in which case it would have the units of spectral radiant sterance [spectral radiance] (watts per square meter steradian micrometer), for example. Thus, Eq. (8-1) may apply to either a sequentially scanning spectrometer or a multiplex spectrometer. A radiometer is a degenerate case of the spectrometer for which Eq. (8-1) contains only one column $(j = 1)$.

The sample mean (the subscript s used to denote "sample" will not be used hereafter) as a function of wavelength λ_j is given by

$$\bar{x}_j(\lambda) = (1/n) \sum_{i=1}^{n} x_{i,j}, \tag{8-2}$$

where x is summed over the index i [down the jth column in (8-1)]. The value of $\bar{x}_j(\lambda)$ should tend toward zero in a well-designed spectrometer. However, coherent microphonics, dc offset, or drift will result in nonzero means. These nonzero values may have the appearance of spectral features or of a continuum background in processed field data.

The mean of the dark-noise output is frequently referred to as the "offset error." The offset error must be subtracted from measured output signals before the data is used to calculate the responsivity, field of view, etc.

The sample standard deviation as a function of wavelength λ_j is given by

$$\sigma_j(\lambda) = \left\{ 1/(n-1) \left[\sum_{i=1}^{n} x_{i,j}^2 - n[\bar{x}_j(\lambda)]^2 \right] \right\}^{1/2}. \tag{8-3}$$

The value of $\sigma_j(\lambda)$ will not necessarily be a constant as a function of wavelength. In general, both $\bar{x}_j(\lambda)$ and $\sigma_j(\lambda)$, may be analyzed in terms of either the output (voltage or current) or the radiation parameter for which the instrument was calibrated. In the latter case, the standard deviation will have the shape of the instrument's inverse spectral responsivity function. In addition, microphonics may increase its value at certain spectral scan positions due to, for example, a rough spot on a drive gear.

Each set of field measurements should include a sufficiently long dark run to provide for an analysis of the noise (as just outlined) under field conditions. The analysis of the dark data will provide a plot of the mean and the standard deviation as a function of wavelength, which may be included as a part of the field report.

8-3   DARK-NOISE DATA PROCESSING

The data may be recorded initially in analog form on magnetic recording tape. Strip chart recordings can then be reproduced as needed. The data can be digitized real time or can be taken from the analog tapes, and codified so that it is possible to identify and process any spectrometer scan included in the calibration run.

The sampling rate must be great enough to preserve the highest frequency content of the noise of the analog output. The sampling theorem [1] states that the highest frequency component is completely determined if sampled at regularly spaced intervals $\frac{1}{2}T$ apart, where T is the period of that component. For example, if the electrical noise bandwidth of the spectrometer output is 100 Hz, the sampling rate must be at least 200 sec^{-1}. Actually, the sampling rate must be greater for practical systems, because the electrical filters are not ideal. Normal practice is to use the factor 5, rather than 2, which for this spectrometer would yield a digitization rate of 500 Hz.

It is necessary to visually inspect each spectrometer scan for irregularities. Any irregular scan should not be used in the analalysis. Such a visual inspection is possible if a catalog of scans is produced. Figure 8-1 shows one page of such a catalog that contains 10 scans of 1-in. deflection. Each scan is identified by the tape number, time code, and file number. Various scans may be selected and identified for processing.

The details given here for dark-noise data processing apply to the analysis of any calibration data such as the field of view, relative spectral responsivity, etc., with respect to recording, digitization rates, and calculation of means and standard deviations.

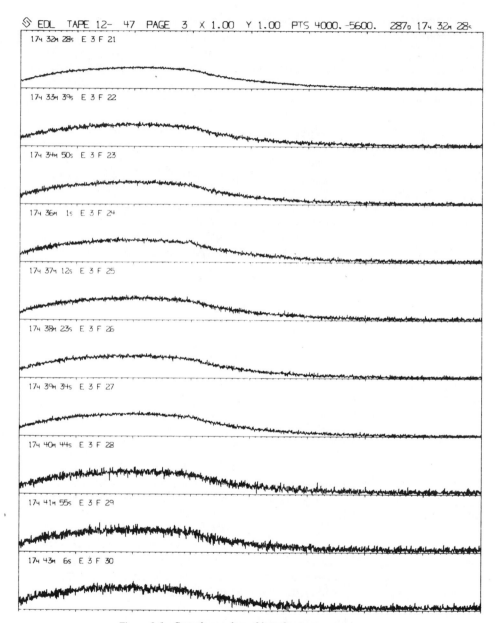

Figure 8-1   Sample catalog of interferometer scans.

8-4 DARK NOISE—AN EXAMPLE

Figure 8-2 gives the dark signal output of a circular variable filter spectrometer. These data represent the output voltage of a $\frac{1}{2}$-sec scan which covers the spectral range of 6 to 24 μm. A visual inspection of the noise scan of Fig. 8-2 reveals a system for which the mean appears to be zero and the noise random.

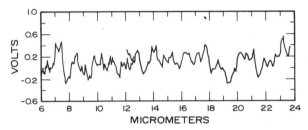

Figure 8-2 Dark-noise scan for a circular variable filter spectrometer—single scan.

Figure 8-3 represents the solution of Eq. (8-2) which gives the mean for each wavelength of 58 noise scans for the same instrument. The noise has decreased as indicated by the change in scale; however, the mean shows that the signal contains both drift and coherent signals. The drift, in the output, is positive and results in an error at 24 μm of nearly 0.1 V. Such drift could be misinterpreted as a continuum background in actual data. The coherent forms which have the appearance of regular positive and negative spikes have been identified as crosstalk resulting from a 20-Hz square wave used as a scan filter position reference signal. This square wave signal is capacitively coupled into the data channel so that it is differentiated, resulting in the alternating positive and negative spikes (the period appears to change at 12.5 μm by a factor of 2 because of a corresponding change in the slope of

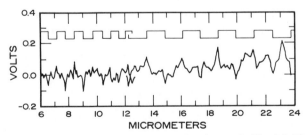

Figure 8-3 The average of 58 dark-noise scans like the one in Fig. 8-2. The square wave shown as an inset at the top is the source of the spikes.

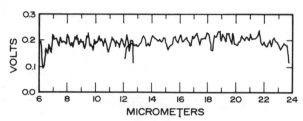

Figure 8-4   The standard deviation as a function of wavelength for 58 dark-noise scans like the one in Fig. 8-2.

the scan position with respect to time). Such coherent noise spikes could be misinterpreted as spectral features in actual data.

Figure 8-4 represents the solution of Eq. (8-3) which gives the standard deviation for the 58 dark scans in terms of voltage (and is quite uniform at 0.2 V) as a function of wavelength. The uncertainty for this channel is 0.6 V for 3σ. Figure 8-2 for the single scan shows a maximum deviation of about 0.45 V. We would not expect the noise voltage to exceed ± 0.2 V some 68% of the time (see Table 7-1).

REFERENCES

1   W. W. Harman, "Principles of the Statistical Theory of Communication," p. 32. McGraw-Hill, New York, 1963.

CHAPTER

IX

Linearity Analysis

9-1 INTRODUCTION

A remote target can be characterized in four nearly independent domains: spatial, spectral, temporal, and polarization. It therefore follows that
the calibration must characterize the sensor responsivity in these four domains. The responsivity of a spectroradiometer is a multivariable functional
relationship between the incident flux and the instrument output:

$$R(\lambda) = R_0 R_s(\lambda)R(\theta, \phi)R_i(\lambda)R(t)R(P), \qquad (9\text{-}1)$$

where R_0 has the units of output per unit flux, $R_s(\lambda)$ the instrument function
(relative to the free spectral range), $R(\theta, \phi)$ the relative spatial response
function (the field of view), $R_i(\lambda)$ the instantaneous relative spectral response
function, $R(t)$ the relative time domain response function, and $R(P)$ the
relative polarization response function. Each of these functions (other than
R_0) is generally peak normalized. All the relative terms in Eq. (9-1) are
defined on the basis of a *linear* response. Thus, where each of these terms
is evaluated theoretically or experimentally throughout this book, it is
assumed that the sensor output signal is linearly related to input flux. It is
necessary to "linearize" the data before evaluating the terms of Eq. (9-1) for

systems that exhibit any degree of nonlinearity. The method by which the data in nonlinear systems is linearized is discussed in the appropriate sections.

It is possible to consider these variables as independent for theoretical considerations and for evaluation in a calibration exercise. Actually they are independent only within certain limits and each must be kept in mind and evaluated to complete the calibration of the sensor.

For example, when measuring the field of view or the spectral bandpass, it is essential that the instrument be scanned through θ and ϕ or λ slowly enough so that the sensor time constant does not distort the shape of the response function. Also, when measuring the spectral bandpass function, the position of the source and its polarization must remain constant; likewise, when measuring the field of view or the polarization response, it is necessary that the source spectral characteristics remain constant.

For linear systems (with zero offset) the functional relation between incident flux Φ and sensor output Γ is given by

$$\Phi = \Gamma(1/R), \tag{9-2}$$

where Φ is the radiant entity of interest. Inherent in R is the complete characterization of the sensor as previously indicated, including the effect of any associated electronic amplifiers, recorders, counters, emulsions, etc., as appropriate.

The term "flux" with the symbol Φ is used three ways (in order to generalize in the theory of calibration), which leads to three responsivities:

$$R_\Phi = \Gamma/\Phi, \tag{9-3}$$

which is the flux responsivity where Φ is power, lumens, quanta per second, etc., and Γ is sensor output;

$$R_E = \Gamma/\Phi_E = \Gamma/E, \tag{9-4}$$

which is the areance [irradiance, illuminance] responsivity; and

$$R_L = \Gamma/\Phi_L = \Gamma/L, \tag{9-5}$$

which is the sterance [radiance, luminance] responsivity. Whenever the term "responsivity" is used without modifiers throughout this book, it could equally well refer to any one of these three responsivities. Likewise, whenever the term "flux" is used without modifiers, it could equally well refer to any one of the radiant entities of interest. Thus, when the flux is given by Eq. (9-2), it could be power, lumens, quanta, areance, or sterance and the units are given, respectively, as flux, sensor output, and output per unit flux.

The major objective of the linearity analysis is to characterize the sensor transfer function. Not all systems are necessarily linear by design or by .

nature. Many detector–preamplifier systems tend to be nonlinear over part of the useful dynamic range. Also, various nonlinear schemes are occasionally used to extend the dynamic range of systems. However, the analysis of the relative responsivity terms of Eq. (9-1) must be obtained with linearized data. It therefore follows that the transfer function must be characterized before these other functions can be determined.

It is the *shape* of the transfer function that must be obtained, rather than its absolute value. It is very convenient to model the relative transfer function as a mathematical equation that describes the entire dynamic range of the sensor output. The absolute transfer function will differ from the relative transfer function by a constant.

Photon detectors measure the rate at which quanta are absorbed, whereas thermal detectors measure the rate at which energy is absorbed. Thus the relative transfer function must be determined as a functional relationship between incident power (or photon rate) and the sensor output. It is essential that the spatial, spectral, polarization, and temporal characteristics of the source remain constant while the magnitude of the power is varied in such a way as to evoke a response throughout the useful dynamic range of the sensor.

The incident power (or photon rate) is proportional to source area for a uniform blackbody source. Therefore the transfer function can be evaluated as a function of source area provided vignetting does not occur. The source area is relatively easily qualified through the use of a precision aperture set. Some ray tracing may be necessary to assure that the image of the aperture on the sensor detector is not larger than the detector.

The incident flux is also proportional to the blackbody temperature according to Planck's equation; however, it is very difficult to qualify the sensor spectral bandpass. Errors in calculating the effective flux introduce an apparent nonlinearity into the data. This problem is avoided by holding the temperature constant and varying the source area as previously explained.

The precision aperture set may not yield a useful range great enough to stimulate the sensor throughout its full dynamic range. In this case a combination of several overlapping aperture sets at different temperatures will provide sufficient overall range. Then the change in flux with temperature can be qualified in the overlap region.

The transfer function of a spectrometer is more conveniently evaluated if a bandpass filter is interposed in the optical path so that for each scan a single "line" is obtained. It is only necessary to read the peak value of that line. A radiometer has a bandpass filter as an inherent property of the sensor. Thus, ideally, the experimental data would consist of a set of output voltages as a function of power, photon rate, or source area.

9-2   GRAPHICAL DISPLAY OF THE TRANSFER FUNCTION

As just indicated, it is convenient to model the transfer function as a mathematical equation. The transfer function is evaluated experimentally by observing the output as a function of the input flux. However, it is necessary to interchange the dependent and independent variables to obtain a functional relationship like that of Eq. (9-1) suitable for use as a calibration equation in which the flux is given as a function of the sensor output.

There are problems that arise in the interpretation of transfer function data as follows:

(1)   A linear system may appear to be nonlinear if vignetting, spectral leakage, or offset errors occur.

(2)   Frequently the dynamic range of linear systems is extended by providing multiple-output channels that differ in gain by some factor. Usually the gain difference factor is set so as to provide some redundancy (or overlap) between channels.

Considerable insight can be obtained about the nature of the transfer function through the use of various graphical displays.

A linear graph is not very useful for extended range systems as most of the points in the data set will appear to plot at the origin of coordinates (near zero). Offset error shows up as a nonzero intercept on each channel. Spectral errors and vignetting " bend " the transfer function. This problem can be overcome by plotting the data on a log–log scale. This is illustrated as follows: The slope of a line (in xy-coordinates), which has the equation

$$y = mx, \qquad (9\text{-}6)$$

is given by $m = (y_2 - y_1)/(x_2 - x_1)$. The slope of Eq. (9-6) m' when plotted on a log–log scale is obtained by taking the log of Eq. (9-6) at the two points to obtain

$$\log y_2 = \log m + \log x_2 \qquad (9\text{-}7)$$

and

$$\log y_1 = \log m + \log x_1. \qquad (9\text{-}8)$$

Subtracting Eq. (9-8) from Eq. (9-7) yields

$$m' \equiv (\log y_2 - \log y_1)/(\log x_2 - \log x_1) = 1, \qquad (9\text{-}9)$$

which illustrates that a linear function with zero offset will plot as a straight line inclined at 45° on a log–log scale. Offset error, spectral errors, and nonlinearity will cause the function to depart from the 45° slope.

The experimental evaluation yields a set of data points (output voltage versus input flux or source area). This data can be evaluated channel by channel (in extended range systems) or as a single set in which the output voltages are referred to one channel by making use of the gain differences. The latter approach is preferred because all the data points should be used in order to obtain a more accurate estimate of the transfer function. However, it is essential that the "offset error" obtained in the "dark-noise analysis" be subtracted from all measurements before the outputs are projected, otherwise the offset error will be amplified by the gain difference. The data can then be graphed on a log–log plot to display the entire dynamic range as a single function. This procedure is illustrated with a 4-channel "linear" spectrometer. The data set is given in Table 9-1 where the output voltage for each

TABLE 9-1

Transfer Function Data

Aperture area[b] (cm^2)	Channel output (Volts)[a]				Relative (Volts)[c]
	G_3	G_2	G_1	G_0	
2.93—4	1.3				1.3
6.30—4	2.3				2.3
1.24—3	4.6				4.6
2.43—3		1.23			12.3
5.08—3		2.15			21.5
9.85—3		3.85			38.5
2.03—2			0.80		80.0
3.99—2			1.60		160
8.17—2			3.40		340
1.63—1				0.65	650
3.27—1				1.05	1050
6.54—1				2.2	2200
1.31—0				3.05	3050

[a] The relative channel gain for the G_n channel is given by 10^n, thus each channel differs by a gain of 10.
[b] The expression 2.93–4 is computer notation for 2.93×10^{-4}.
[c] Referred to G_3.

of 4 channels is given as a function of source area at a fixed blackbody temperature for a precision aperture set that varies in steps of 2. The output voltages are referred to the level of the high-gain channel (G_3) by multiplying each point by the relative channel gain.

A linear graph of the data listed in Table 9-1 is given in Fig. 9-1. The data appears to be nonlinear; however, the lower 5 data points are too near the

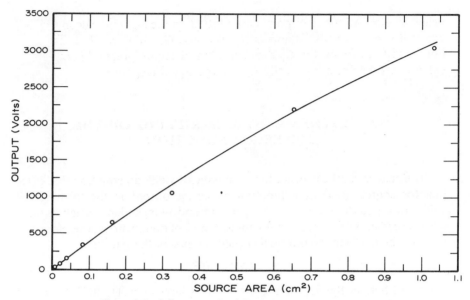

Figure 9-1   A linear graph of the data listed in Table 9-1.

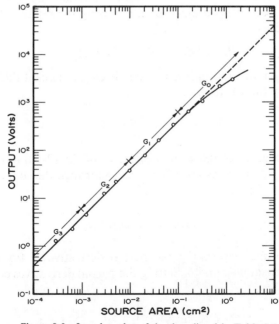

Figure 9-2   Log–log plot of the data listed in Table 9-1.

origin to plot. All the data points are plotted on the log–log scale in Fig. 9-2 where it is obvious that channels G_3, G_2, and G_1 are linear, but that some nonlinearity exists on the G_0 channel. Most detector–preamplifiers tend to exhibit this type of nonlinearity for a large-signal response.

9-3   MATHEMATICAL MODELING OF THE TRANSFER FUNCTION

The analysis of the transfer function data set yields an equation that gives flux (or source area) as a function of sensor output in the form of the calibration equation where the dependent and independent variables have been reversed. All systems exhibit some degree of nonlinearity; therefore, the form of the mathematical model should be that of the parabola

$$A \doteq a\Gamma + b\Gamma^2, \tag{9-10}$$

where A becomes the flux of interest (or the source area A) and Γ the sensor output. The constant a corresponds to the inverse responsivity $(1/R)$ of Eq. (9-2) while the constant b is a measure of the degree of nonlinearity.

The data set is used to obtain a "best fit" equation based on the form of Eq. (9-10). The best fit is obtained by minimizing the relative difference D_i which is defined as

$$D_i = [A_i - (a\Gamma_i + b\Gamma_i^2)]/A_i. \tag{9-11}$$

The relative difference multiplied by 100 yields the percent difference. The error is given by the variance

$$\sigma^2 = \sum_{i=1}^{n} D_i^2/(n-2) \tag{9-12}$$

which gives a measure of the quality of the fit of Eq. (9-10) to the data set. The quantity $n - 2$ is a constant which passes through the summation; thus the quantity that should be minimized is

$$\text{error} = \sum_{i=1}^{n} D_i^2. \tag{9-13}$$

This is accomplished by taking the partial derivative of Eq. (9-13) with respect to the constants a and b. Setting the partial derivatives equal to zero yields the equations

$$\sum_{i=1}^{n} (\Gamma_i^2/A_i^2)a + \sum_{i=1}^{n} (\Gamma_i^3/A_i^2)b = \sum_{i=1}^{n} \Gamma_i/A_i \tag{9-14}$$

and

$$\sum_{i=1}^{n} (\Gamma_i^3/A_i^2)a + \sum_{i=1}^{n} (\Gamma^4/A^2)b = \sum_{i=1}^{n} \Gamma^2/A_i. \qquad (9\text{-}15)$$

Equations (9-14) and (9-15) can be solved simultaneously for the constants a and b for the best-fit equation for the transfer function data set. The standard deviation can also be found which gives a measure of the quality of the fit.

Equations (9-14) and (9-15) can be solved by determinants, for example, as follows: Values are obtained for each of the coefficients based on the points in the data set where Γ_i corresponds to the voltage obtained in a measurement and A_i the corresponding flux or source area. These coefficients are

$$\sum_{i=1}^{n} \Gamma_i/A_i = A, \qquad (9\text{-}16)$$

$$\sum_{i=1}^{n} \Gamma_i^2/A_i = B, \qquad (9\text{-}17)$$

$$\sum_{i=1}^{n} \Gamma_i^2/A_i^2 = C, \qquad (9\text{-}18)$$

$$\sum_{i=1}^{n} \Gamma_i^3/A_i^2 = D, \qquad (9\text{-}19)$$

$$\sum_{i=1}^{n} \Gamma_i^4/A_i^2 = E, \qquad (9\text{-}20)$$

By determinants, the value of a is given by

$$a = (AE - BD)/(CE - D^2) \qquad (9\text{-}21)$$

and

$$b = (CB - DA)/(CE - D^2). \qquad (9\text{-}22)$$

Thus the transfer function based on the form of Eq. (9-10) is developed. This form is recommended for all so-called "linear" systems. The percentage of full-scale nonlinearity is calculated by taking the ratio of $b\Gamma^2$ to $a\Gamma$ in Eq. (9-10) to provide a measure of the degree of nonlinearity of the system. Calculated as a percentage, this ratio will be very small for truly linear systems so that the second-order term can be neglected.

Equation (9-10) differs from the absolute transfer function (the calibration equation) by a constant, and is "anchored" by another set of data points obtained in the *absolute calibration*.

The quality of fit of the data to the transfer function is

$$\sigma = \left(\frac{1}{n-2}\sum_{i=1}^{n}D_i{}^2\right)^{1/2} \times 100\%, \qquad (9\text{-}23)$$

which is the standard deviation expressed as a percentage.

9-4 NONLINEAR SYSTEMS

Various nonlinear schemes are occasionally used to extend the dynamic range of systems when the cost of telemetry is too high to permit the use of multiple linear output channels. The dynamic range of a typical telemetry channel is 100 which corresponds to 40 dB since

$$dB = 20 \log_{10}(V_m/V_n), \qquad (9\text{-}24)$$

where V_m represents the maximum (or full-scale) channel output voltage, V_n the channel noise voltage, and the ratio V_m/V_n is typically 100. Many detectors–transducers can be operated linearly over a dynamic range of 10^4 to 10^6.

Two techniques are currently in use to achieve more than the normal 40-dB dynamic range of a single telemetry channel: (1) the logarithmic transfer function and (2) the so-called "dogleg" transfer function.

9-4-1 Logarithmic Systems

There are two types of logarithmic sensors in use. The first, considered here, makes use of an amplifier that takes the $\log_{10}$ of the preamplifier dc output voltage V_p which is given by

$$V_p = V_s + V_0 \quad [\text{V}], \qquad (9\text{-}25)$$

where V_0 is the dc offset voltage and V_s the signal voltage given by

$$V_s = R\Phi \quad [\text{V}] \qquad (9\text{-}26)$$

with R the responsivity in units of volts per flux. Thus the output voltage V_L of the log–amp is given by

$$V_L = \log(R\Phi + V_0) \quad [\text{V}]. \qquad (9\text{-}27)$$

The calibration equation is obtained by solving Eq. (9-27) for Φ, which, by the definition of logarithms, is

$$\Phi = (1/R)[10^{V_L} - V_0] \quad [\Phi]. \qquad (9\text{-}28)$$

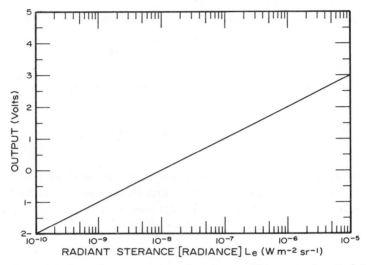

Figure 9-3 Input flux versus voltage for a log amplifier system, $\Phi = 10^{-8}\ 10^{V}$.

Figure 9-3 illustrates the log system response function for the case where the inverse responsivity has a magnitude of 10^{-8} and V_0 is zero. This system has the disadvantages that the intercept is dependent on the offset voltage and that negative voltages can result from low flux levels.

Another type of logarithmic system that is occasionally used with multiplier phototube (pm tube) detectors (which behave as a current source) is one that incorporates a solid-state diode. Such a diode functions as a logarithmic current to voltage converter.

The current i through a diode (at a constant temperature) is given by [1]

$$i = I_0(e^{V_d/a} - 1)\quad [\text{A}]\tag{9-29}$$

where I_0 is called the *reverse saturation current*. When the diode is reverse-biased with a relatively large negative diode voltage V_d, the current is $i = -I_0$ and the reverse saturation current is independent of the applied voltage. When the voltage V_d is positive and relatively large, the unity term can be neglected in Eq. (9-29) and the current

$$i = I_0 e^{V_d/a}\tag{9-30}$$

increases exponentially with voltage.

The pm tube output current i consists of the sum of the signal current i_s and a dark current (the current with zero incident flux) i_d; thus

$$i = i_s + I_d\quad [\text{A}],\tag{9-31}$$

where the signal current is given by

$$i_s = \Phi R \quad [A] \tag{9-32}$$

with R the current resposivity in units of amperes per unit flux.

These equations can be combined to give

$$\Phi R + I_d = I_0 e^{V_d/a} \quad [A]. \tag{9-33}$$

Solving Eq. (9-33) for the output voltage yields

$$V_d = a \ln[(\Phi R + I_d)/I_0] \quad [V], \tag{9-34}$$

which shows that the voltage cannot go to zero because of the dark current.

However, zero output can be obtained by introducing an offset voltage in series with the output which has a magnitude of

$$V_0 = -a \ln(I_d/I_0) \quad [V]. \tag{9-35}$$

The output voltage for the sum of V_d and V_0 is given by

$$V = a \ln[(\Phi R + I_d)/I_d] \quad [V], \tag{9-36}$$

which yields $V = 0$ for $\Phi = 0$.

Solving Eq. (9-36) for Φ yields the calibration equation for the diode logarithmic transfer function

$$\Phi = (I_d/R)(e^{V/a} - 1) \quad [\Phi], \tag{9-37}$$

which is illustrated in Fig. 9-4 for a logarithmic photometer.

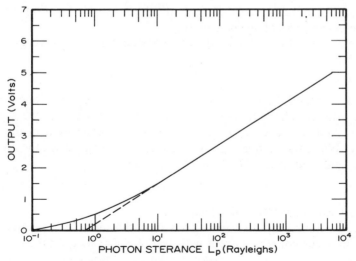

Figure 9-4   Input flux versus voltage for a pm tube logarithmic system. The solid curve is $L_p = 0.64 \, (e^{1.83V} - 1)$ and the dashed curve is $L_p = 0.64 e^{1.83V}$.

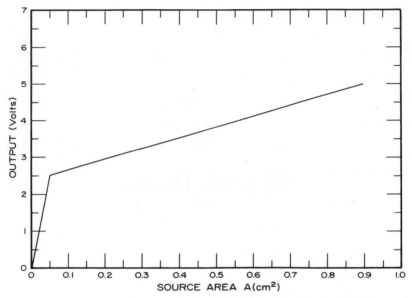

Figure 9-5 Illustration of the dog leg transfer function where the horizontal part of the graph is $A = (1/2.890)V - 0.83$ and the vertical part is $A = (1/50)V$.

9-4-2 The Dogleg Transfer Function

The dogleg transfer function is essentially two channels compressed and stacked together into a single channel as illustrated in Fig. 9-5. The response for the two halves is linear and is given by

$$\Phi = (1/R)\Gamma \qquad\qquad (9\text{-}38)$$

for the lower half and

$$\Phi = (1/R)\Gamma + \Phi_0 \qquad\qquad (9\text{-}39)$$

for the upper half, where the inverse responsivity $1/R$ takes on different values for each range and Φ_0 is the zero intercept for the upper channel. The difficulties with this system are that the signal-to-noise ratio or precision is decreased by the factor 2, and the value of Φ_0 is difficult to determine accurately and may not be stable except in carefully designed systems.

REFERENCES

1   J. Millman and C. C. Halkias, "Electronic Fundamentals and Applications for Engineers and Scientists," p. 26. McGraw-Hill, New York, 1976.

CHAPTER

X

Spatial Purity

10-1 INTRODUCTION

The sensor aperture is bombarded with unwanted flux from sources outside the field of view. The sensor output for a spatially pure measurement is a function of the flux originating from the target (within the field of view), and is completely independent of any flux originating from outside this region. Thus the calibration of the spatial response (or the field of view) of a sensor is considered, in this book, as a problem of spatial purity.

All practical sensors exhibit some off-axis response and large measurement errors can occur. The off-axis response is a problem of optical baffling in optical sensor design. The degree to which the off-axis response must be characterized depends on the nature of the target and the background in which it is immersed. However, it may be necessary to evaluate the off-axis response to many orders of magnitude below the on-axis (in-field) response; this can be a very difficult aspect of sensor calibration.

The solar coronagraph is an example of an electrooptical sensor in which the radiation of the sun is effectively blocked so that the brighter parts of the corona can be measured at the experimenter's convenience, rather than waiting for a solar eclipse. Other examples include infrared antiaircraft missile trackers that must reject the sun, and various radiometers and spectrometers designed to measure atmospheric emission species in the presence of the sun, moon, and earth. What constitutes one person's source may be another person's background; however, the problem of the off-axis source is one that must be considered in any measurement program.

The purpose of this chapter is to deal with spatial purity. The problems of spatial purity arise from instrument response that exhibits nonideal spatial responsivity and source spatial distribution that is nonuniform or unknown (except for the case of a point source).

Often the spatial distribution of a field source radiant energy is beyond the control of the observer, and in fact, may even be unknown. However, the instrument spatial response (field of view) is somewhat amenable to design. The emphasis of this chapter is therefore placed on the definition of the field of view (both ideal and nonideal), the errors associated with the measurement of nonuniform sources, and with incomplete characterization of the sensor field of view. The details of the field-of-view calibration technique, equipment, and data processing are given in the succeeding chapters.

10-2 FIELD OF VIEW

The objective of a field measurement of an extended source using an electrooptical sensor, with the optical axis oriented along a given direction, is to obtain the total flux in a narrow region centered about the optical axis as shown in Fig. 10-1. The source spatial distribution of flux as a function of the spherical coordinates θ and ϕ for an extended source is given by $\Phi(\theta, \phi)$ over the surface of a hemisphere that is oriented with the hemispherical Z-axis along the instrument optical axis. Then the total flux is given by the integral of $\Phi(\theta, \phi)$ over the projected field of view Ω of the sensor

$$\Phi_T = \int_\Omega \Phi(\theta, \phi) \, d\Omega \quad [\Phi], \tag{10-1}$$

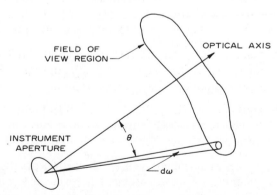

Figure 10-1 Illustration of a sensor field of view. [Adapted from *Proc. Soc. Photo-Opt. Instrum. Eng.—Infrared Technol. II* **95**, 219 (1976).]

where Ω is given by

$$\Omega = \int \cos \theta \, d\omega \quad [\text{sr}] \tag{10-2}$$

with $d\omega$ the incremental element of solid angle and θ the polar angle between the optical axis and the element $d\omega$ as shown in Figure 10-1.

The actual sensor output Γ in response to $\Phi(\theta, \phi)$ is given by the integral of the product of $\Phi(\theta, \phi)$ with the sensor responsivity $R_0 R(\theta, \phi)$:

$$\Gamma = R_0 \int_{(\text{hemisph})} R(\theta, \phi)\Phi(\theta, \phi) \, d\Omega \quad [\text{output}], \tag{10-3}$$

where the integration is carried out over the whole hemisphere and R_0 the peak responsivity in the direction of the optical axis (with $R(\lambda)$, $R(P)$, and $R(t)$ constant).

From this the measured target flux is obtained by the relation

$$\Phi_m = \Gamma/R_0 = \int_{(\text{hemisph})} R(\theta, \phi)\Phi(\theta, \phi) \, d\Omega \quad [\Phi], \tag{10-4}$$

which, based on Eq. (10-1), is equal to Φ_T only if

$$\int_{(\text{hemisph})} R(\theta, \phi)\Phi(\theta, \phi) \, d\Omega = \int_{\Omega} \Phi(\theta, \phi) \, d\Omega \quad [\Phi]. \tag{10-5}$$

Unfortunately, the equality in Eq. (10-5) is valid under only two conditions, neither of which can be perfectly realized in practice.

(1)   The equality in Eq. (10-5) is valid for an ideal instrument relative spatial response function $R(\theta, \phi)$, which is defined as one that has unity response over the region of the instrument field of view and is zero elsewhere. In this case, since $R(\theta, \phi)$ is a unity constant which passes through the integral and the limits change to Ω, then Eq. (10-5) is identically true.

Such an ideal response function will always yield the total field flux Φ_T in the measurement regardless of the shape of $\Phi(\theta, \phi)$; consequently great effort is expended to achieve practical systems that approach the ideal.

(2)   The equality in Eq. (10-5) is also valid for the case where $\Phi(\theta, \phi) = \Phi_0$ (a constant) which requires that it be either a uniform extended source or a point source (approaching zero extent).

Case (1), the ideal instrument, is not achieved in practical instruments; therefore, the meaning of a measurement is given in terms of case (2) as follows: " Provided that $\Phi(\theta, \phi)$ is spatially uniform or a point source, it has a magnitude of Φ_T as calculated by Eq. (10-4)."

Under these practical conditions [case (2)], Eq. (10-5) becomes

$$\int_{(\text{hemisph})} R(\theta, \phi) \, d\Omega = \int_{\Omega} d\Omega = \Omega \quad [\text{sr}]. \tag{10-6}$$

The right-hand side of Eq. (10-6) is an expression of the "ideal" field of view where the limits on the integral define the field of view. The left-hand side of Eq. (10-6) is an expression of the nonideal field of view.

10-3   THE IDEAL FIELD OF VIEW

Equation (10-6) is the basis for a definition of both the ideal and nonideal field of view. The ideal field of view is given by

$$\Omega = \int_{\Omega} d\Omega = \int_{\Omega} \cos \theta \, d\omega \tag{10-7}$$

which may be visualized in terms of a circularly symmetrical field of view as illustrated in Fig. 10-2. The circularly symmetrical case is convenient to

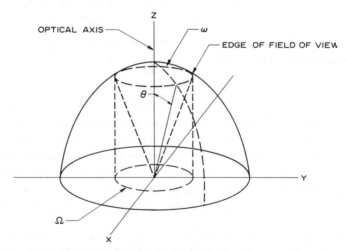

Figure 10-2   Illustration of the ideal field of view.

illustrate the computation of the solid angle field of view because the relative response function reduces to a one-dimensional function $R(\theta)$ and because it is typical of many practical systems.

The ideal field of view is one for which $R(\theta)$ has a value of unity for $\theta = 0$ to $\theta = \Theta$, and is zero elsewhere. Thus Eq. (10-7) can be written, for the circularly symmetric case, as

$$\Omega = \int_0^{2\pi} d\phi \int_0^{\Theta} \cos\theta \sin\theta \, d\theta = \pi \sin^2 \Theta \quad [\text{sr}]. \tag{10-8}$$

The projected field of view must always be used in calculations based on the invariance of sterance [radiance, luminance] or in calculations of sensor throughput. Therefore, whenever an instrument field of view is given with the symbol Ω it is understood to be the *projected* field of view.

The field of view ω is given by

$$\omega = \int_0^{2\pi} d\phi \int_0^{\Theta} \sin\theta \, d\theta = 2\pi(1 - \cos\Theta) \quad [\text{sr}] \tag{10-9}$$

which has the same numerical value as Ω to within 1% when Θ is less than 10°. The angle Θ in Eqs. (10-8) and (10-9) is the half-angle edge of the ideal field of view, and is the edge of the right circular cone that defines the field of view in Fig. 10-2.

10-4   THE NONIDEAL FIELD OF VIEW

The nonideal field of view is also defined by Eq. (10-6), which for the circularly symmetric case is given by

$$\Omega = \int_{(\text{hemisph})} R(\theta) \, d\Omega \quad [\text{sr}], \tag{10-10}$$

where $R(\theta)$ is the normalized relative response which is obtained as the measured off-axis response to a point source. Equation (10-10) is approximated by incrementing $R(\theta)$ as shown in Fig. 10-3. Each increment of Ω is calculated by weighing $\pi \sin^2 \Theta_i$ [Eq. (10-8)] with the corresponding midpoint value of R_i for each increment to yield

$$\Delta\Omega_i = \pi R_i(\sin^2\theta_i - \sin^2\theta_{i-1}) \quad [\text{sr}]. \tag{10-11}$$

The effectiveness of each incremental solid angle $\Delta\Omega_i$ to produce an output signal depends on the value of R_i; thus the resultant total solid angle, the sum of all weighted increments, might be termed *effective*, and is given by

$$\Omega_{\text{eff}} = \pi \sum_{i=1}^{n} R_i(\sin^2\theta_i - \sin^2\theta_{i-1}) \quad [\text{sr}]. \tag{10-12}$$

However, since $R(\theta)$ is normalized to the peak value of the off-axis response function, it is also appropriate to refer to the resultant solid angle as the *peak*

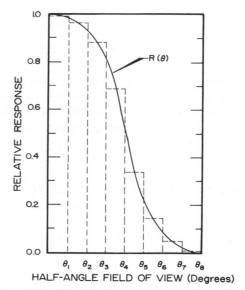

Figure 10-3 The off-axis response to a point source.

normalized solid angle field of view. In any calibration it is important to specify the type of normalization used.

The half-angle field of view Θ of Figure 10-2 for the ideal field of view that is equivalent to Ω_{eff} is obtained by combining Eqs. (10-8) and (10-12) and solving for Θ:

$$\Theta_{eff} = \arcsin \left[\sum_{i=1}^{n} R_i(\sin^2 \theta_i - \sin^2 \theta_{i-1}) \right]^{1/2}, \qquad (10-13)$$

where Θ_{eff} is the half-width of an *equivalent* ideal response function that has $R(\theta) = 1$ for $\theta = 0$ to $\theta = \Theta$, and is zero elsewhere. The calculation of Θ_{eff} in Eq. (10-13) is equivalent to computing the width of an ideal $R(\theta)$ that has the same area under the normalized curve as does the nonideal function $R(\theta)$.

An accurate evaluation of Θ_{eff} or Ω_{eff} requires that the off-axis values of $R(\theta)$ be evaluated and summed in Eqs. (10-12) and (10-13) for many orders of magnitude below the on-axis response.

For many systems, the response function $R(\theta)$ is very nearly *Gaussian* or *natural*. The Gaussian function is given by

$$R(x) = \exp(-x^2/1.288\sigma^2), \qquad (10-14)$$

where $R(x)$ has the value of unity for $x = 0$, zero for $x = \infty$, and σ is the half-width corresponding to $R(x) = 0.460$. The integral of $R(x)$ is equal to the half-width of an equivalent ideal response function, that is,

$$\int_0^\infty \exp(-x^2/1.288\sigma^2)\,dx = \sigma \qquad (10\text{-}15)$$

to within 1%. Thus the edge of the ideal equivalent response function passes through the 0.460 point on the natural function. Common practice is to specify the half-power point $R(x) = 0.50$ as the equivalent ideal width as a first-order approximation for nonideal functions that are well behaved, i.e., that are Gaussian.

10-5   ERRORS ASSOCIATED WITH NONIDEAL FIELD OF VIEW

The preceding conditions for which the exact value of source flux can be obtained from a field measurement are:

(1)   The instrument field of view is ideal.
(2)   The source is spatially uniform.
(3)   The source is a point source.

The measurement of either an extended source or a point source may exhibit large errors in the presence of high-intensity, off-axis sources.

Corrections can be made for the shape of the spatial distribution, provided it is known.

The spatial distribution of the flux can be expressed as

$$\Phi(\theta, \phi) = k\Phi_r(\theta, \phi) \qquad (10\text{-}16)$$

where $\Phi_r(\theta, \phi)$ is the relative spatial distribution of the source flux (or the shape of the distribution) and the coefficient k gives the absolute units. The sensor output can be expressed in terms of Eqs. (10-3) and (10-16) as

$$\Gamma = R_0 k \int_{(\text{hemisph})} \Phi_r(\theta, \phi)R(\theta, \phi)\,d\Omega, \qquad (10\text{-}17)$$

from which k is obtained as

$$k = \Gamma \Big/ \Big(R_0 \int_{(\text{hemisph})} \Phi_r(\theta, \phi)R(\theta, \phi)\,d\Omega \Big). \qquad (10\text{-}18)$$

The total flux is then obtained as

$$\Phi = k \int_\Omega \Phi_r(\theta, \phi)\,d\Omega, \qquad (10\text{-}19)$$

where Ω defines any spatial region that satisfies the measurement objectives and for which the relative spatial distribution of the source is available.

The solution of Eqs. (10-18) and (10-19) requires that $\Phi_r(\theta, \phi)$ be known either from measurement or from theory. It is also necessary to know $R(\theta, \phi)$, which is an important and difficult aspect of sensor calibration. If such a correction is not made, the data should be reported as *peak normalized*. In any case it is important to report the function $R(\theta, \phi)$ so that the reader can apply his own corrections.

The normalized flux Φ_m obtained by Eq. (10-4) differs from the total flux Φ_T of Eq. (10-1) in a way that depends on the distribution of both $R(\theta, \phi)$ and $\Phi(\theta, \phi)$. The difference (error for uncorrected data) can be quite large when, for example, a nonuniform source is measured with a nonuniform field of view. The errors can be classified under two types: *near field* and *far field*.

The near field is defined as the sensor response near the on-axis response. The near-field evaluation provides for a direct comparison with the ideal field of view by which it is characterized, while the far-field evaluation gives a measure of the ability of the system to block out off-axis (out-of-field) sources, and must be evaluated to many orders of magnitude below the on-axis response.

Near-field errors result from having bright regions of the source coincide with below-peak response regions of the field of view, while far-field errors result from having very intense off-axis (out-of-field) sources that are capable of producing a significant contribution to the sensor output even though the sensor spatial response to that source may be many orders of magnitude below the on-axis response.

XI

Field of View Calibration

11-1 INTRODUCTION

The field-of-view response function $R(\theta, \phi)$ is defined as the normalized response to a point source. A distant small area source can be used to simulate a point source; however, a small area source used in conjunction with a collimator is more convenient.

A typical facility for calibrating an optical field of view is illustrated in Fig. 11-1. In this figure, a cylindrical sensor is shown. The sensor is mounted in a cradle on a calibrated turntable so that it may be rotated in azimuthal angle about its aperture, or in elevation angle. The turntable is also adjustable in height. A collimator, blackbody source, and aperture plate are shown mounted on an optical bench to provide for alignment and for a convenient method to mount auxiliary optical devices such as the reticle shown.

The evaluation of the field of view of a spectrometer may be facilitated by interposing a narrow-band optical filter, such as a bandpass interference filter, between the blackbody source and the instrument aperture so that it is only necessary to read the peak value of the resulting spectral "line." However, it is necessary to step the instrument through the field since the instrument responds only once for each spectral scan. This aspect of the calibration of the field of view is very inconvenient and greatly complicates the data reduction. Fortunately, this problem does not occur in a nonscanning radiometer.

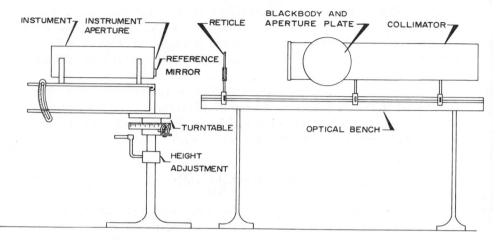

Figure 11-1   Point source field-of-view calibration facility.

It is sometimes possible to stop the spectrometer at some suitable wavelength and take the output signal from an internal test point (for example, the preamplifier output); then the spectrometer functions as a simple radiometer. The instrument can then be geometrically "scanned" to yield a continuous output function of the azimuthal angle or of the elevation angle, as appropriate.

The output data, the sensor response to the simulated point source as a function of θ and ϕ, must be interpreted before it can be applied directly to the problem of the characterization of the sensor. The rule of good performance (that the calibration should be conducted under conditions which reproduce, as completely as possible, the conditions under which the measurements are to be made) often can be compiled with as far as the spatial parameters are concerned. For example, an extended area source can be used for extended source measurements or, conversely, a small area source can be used for point source measurements. The field-of-view data does not enter directly into the determination of the sensor responsivity when this ideal is observed. However, where a small area source calibration must be extrapolated for full-field, extended source measurements, or vice versa, the magnitude of the projected solid angle Ω must be known.

The normalized flux Φ_N, as measured with a practical sensor (nonideal field of view), differs from the total flux Φ_T. Corrections can be made in the measured flux provided that $\Phi(\theta, \phi)$ and $R(\theta, \phi)$ are known. Thus the shape of the $R(\theta, \phi)$ function must be evaluated if corrections are to be made. The extent of the difference (the error) can be estimated from the shape of the field of view even when the distribution of the flux $\Phi(\theta, \phi)$ is not known.

A consideration of the type of errors that occur in nonideal sensors leads to the distinction being made in the *near-field* and the *far-field* response of the field of view. This distinction is also a practical one inasmuch as the experimental evaluation, or calibration, of the sensor field of view falls into the same two categories.

The next section contains a description of the practical aspects of field-of-view calibration including considerations of resolution, optical alignment, and the evaluation of the near-field and far-field spatial response. Included are detailed examples of the calibration of a sensor field of view.

11-2 RESOLUTION

The amount of detail contained in the field-of-view data depends on how well the small-area source and collimator simulate a true point source (see Figure 11-1). The "shape" of the field-of-view response is resolved to the

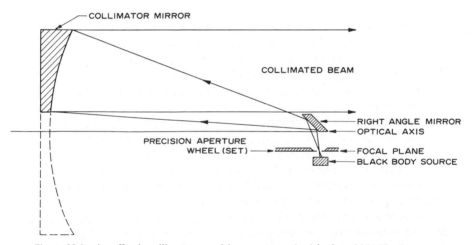

Figure 11-2 An off-axis collimator, precision aperture wheel (set), and blackbody source.

level of the divergence $\Delta\theta$ of the collimated beam. If a small area source is placed at the collimator focus as shown in Fig. 11-2, the resultant rays will have a divergence given by

$$\text{arc } \tan(D_s/2F) = \Delta\phi, \tag{11-1}$$

where D_s is the source diameter and F the collimator focal length.

A set of precision apertures is frequently used in conjunction with a uniform blackbody source to control both the source area and the divergence.

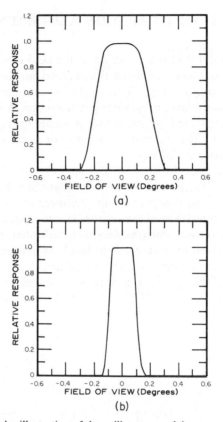

Figure 11-3   An illustration of the collimator resolving power; (a) and (b).

The collimator resolution must be kept small compared to the detail characteristic of the sensor field of view. For example, a photometer with a circularly symmetric field of view of 3.50 mrad (0.10° half-angle) was evaluated in a set up like that of Figs. 11-1 and 11-2. The output of the sensor, as a function of azimuthal angle, is illustrated in Fig. 11-3 as a normalized cross-sectional field of view. Part (a) shows the result of using a relatively large aperture that produced a beam divergence of 0.32°, while part (b) resulted from a relatively small aperture that yielded a divergence of 0.027°. Figure 11-3a provides only an evaluation of the collimator resolution, while Fig. 11-3b provides the desired photometer field of view. Part (a) illustrates how an improperly designed experiment yields instrument attributes (the collimator in this case) rather than the desired target attribute (the sensor in this case).

11-3 OPTICAL AXIS ALIGNMENT

It is often required that the direction of the field-of-view optical axis be evaluated with respect to some physical surface to facilitate alignment of the sensor optical axis on the target. This might be a mounting flange surface, the entrance aperture plane, or a boresight telescope. A simple yet effective technique is accomplished by mounting a small mirror adjacent to the entrance aperture so the normal to the mirror surface is approximately parallel to the sensor optical axis.

An evaluation of the direction of the optical axis of the sensor can be qualified in terms of the normal to the mirror surface. The mirror can then be used as a reference surface permitting alignment of the sensor optical axis independent of the operation of the sensor. This is especially important with difficult-to-operate sensors such as sensitive multiplier–phototube systems that may be overdriven by ambient light levels or of cryogenic systems that are subject to vacuum or background limitations.

The direction of the optical axis may be qualified with the point source facility as illustrated in Figs. 11-1 and 11-4. A fixed reticle with a cross-

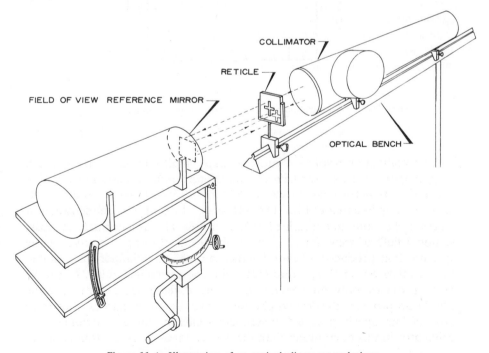

Figure 11-4 Illustration of an optical alignment technique.

shaped opening is placed between the collimator and the sensor-mounted reference mirror. The turntable height is adjusted so the light rays that pass through the cross from the collimator are reflected back from the mirror upon the reticle, as illustrated in Fig. 11-4. The turntable is adjusted in vertical and azimuthal angle until the reflected cross falls directly back on the opening in the reticle. This provides coalignment of the normal to the mirror with the collimator output rays. The turntable fiducial (origin or reference zero) is set to correspond to that position and then the field of view mapped against that reference zero.

11-4   OFF-AXIS REJECTION

The experimental evaluation of the off-axis rejection characteristics of well-baffled systems is extremely difficult. This is because of scattering off the sensor aperture back onto the walls of the room in which the measurements are being made. The scattering levels may be sufficient to mask the true characteristic of the off-axis response even when the walls and the equipment surfaces have been blackened.

The scattering typical of a calibration facility (illustrated in Fig. 11-5) can be approximated by assuming the walls to behave as perfectly diffuse Lambertian surfaces scattering a fraction δ of the incident energy. The ratio β of flux density scattered back into the sensor aperture to that of the collimator beam is given by

$$\beta = (\delta \Omega A \cos \phi \cos \theta)/\pi^2 s^2, \tag{11-2}$$

where θ is the angle between the optical axis of the sensor and the collimator axis, ϕ the angle between the sensor axis and the normal to the wall, s the distance between the sensor aperture and the wall, and A the collector area.

For example, the value of Eq. (11-2) is approximately 10^{-7} for a typical facility where δ is 0.05, Ω 0.006 sr (5° full-angle field of view), $\phi = \theta = 45°$, and s is 2 m.

Notice that the scattering is proportional to the sensor solid angle field of view Ω. The value of β obtained would, therefore, depend on the particular sensor being calibrated.

This calculation is based on a facility treated with a diffuse black surface preparation with a rectangular shape. An alternative technique is to make use of a specular (reflective) black and a special geometrical shape to reflect the radiation away from the sensor. Such a facility was designed [1] in which the aerosol scattering is also reduced by locating the chamber within a class-100, clean-room environment. The floor plan of this facility, contained within a 3.25 × 6.1 m laboratory room, is shown in Fig. 11-6. The design

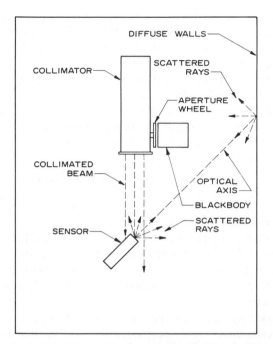

Figure 11-5 Floor plan of a typical calibration facility.

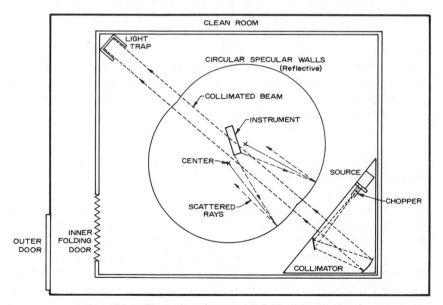

Figure 11-6 Floor plan of a specular field of view facility.

incorporates a circular shape which has two centers with the sensor placed between the centers. The energy scattered from the sensor aperture is reflected off the circular (cylindrical) walls *away* from the sensor in this design.

Surface scattering and aerosol scattering are sufficiently reduced so that the limiting background results from Rayleigh (molecular) scattering. In this case the scattering is proportional to the scattering volume which is also a function of the solid-angle field of view Ω of the sensor.

Measurements of the ratio of on-axis to off-axis response β was measured at 1×10^{-9} for a 5° full-angle sensor at $\theta = 40°$—an improvement of two orders of magnitude over the diffuse black, rectangular facility previously mentioned.

The off-axis rejection, or the far-field response to a point source, is very difficult to evaluate as previously indicated. For this reason it is not practical to map the entire field, but rather to examine a cross-sectional response rather carefully. This might be accomplished with a facility like that just described as follows: The sensor is physically aligned so that the optical axis coincides with the axis of the collimator. The source, which might be a laser, blackbody, or ribbon lamp, is adjusted in intensity until the sensor output is nearly saturated on its lowest gain output channel. The intensity of the source may be adjusted either by controlling the source or by interposing neutral density filters in the optical path.

The sensor is then rotated off-axis, while the position reference signal and sensor output signal are simultaneously recorded, until the output signal falls nearly to the noise level on the highest gain output channel. At this point the intensity of the source must be increased, either by increasing the electrical power to the source or by removing some of the neutral density filters (without moving the sensor), until the output signal is nearly saturated on its lowest gain channel again. This requires that the source intensity be increased by a value roughly equal to the dynamic range of the sensor.

The ratio of the final intensity of the source to its initial value can be determined from the linearized output signal of the sensor. It may also be possible to determine this ratio based on previous calibrations of the source intensity or the neutral density filters.

This process can be repeated as many times as necessary to define the off-axis response until limited by Rayleigh scattering. The data from each run must be "unfolded" in order to provide a continuous output as a function of scan position. This is accomplished by multiplying each set of data by the appropriate source intensity ratio. This method may suffer from accumulative errors in determining the source ratios.

There exists a dynamic range limit to this folded method of evaluating the off-axis response that depends on the dynamic range of both the source and the sensor. For example, a 4-channel system with a gain difference of 10 between channels has a dynamic range of 10^5 (considering each channel to have a dynamic range of 100). In this case the off-axis response could be evaluated to 10^{-5} of the on-axis response in the first step. The off-axis response could be evaluated to 10^{-10} of the on-axis response, if the source intensity could be increased by 5 orders of magnitude in a second folded step.

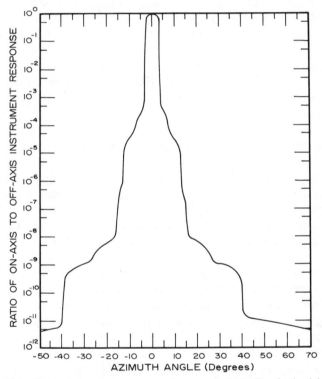

Figure 11-7   Field-of-view measurement of a sun shade baffle obtained in a two-step procedure.

A two-step process permitted measurements down to less than 10^{-11} of the on-axis response in the visible upon a multiplier–phototube system fitted with a sunshade baffle as shown in Fig. 11-7. The shape of the cross-sectional far-field response of Fig. 11-7 contains breakpoints at 2.5, 12, 16, 26, and 40° off-axis that are characteristic of the baffle design. They result from the illumination of various stops and baffle edges that scatter into the sensor aperture.

Powerful CO_2 (beam-expanded) lasers are often used at 10 μm with a bank of neutral density filters to effect off-axis measurements in the IR. The use of neutral density filters is fraught with difficulties because their behavior is difficult to predict, especially when used in physical arrangements where multiple reflections occur.

Theoretical ray-trace techniques are available that can be used to predict the off-axis rejection of optical systems [2, 3]. The reliability of such calculations can be experimentally verified as previously described. Such calculations permit extending the measurements and thus defining the off-axis rejection to whatever degree may be required.

11-5   FIELD-OF-VIEW ANALYSIS

The practical problems associated with the experimental evaluation of the field of view leads to a classification of measurements referred to as *near field* and *far field*. The near field is defined as the linearized and normalized response to a point source at angles *near* the optical axis, as evaluated in detail to about 1% of the on-axis response. The near field is graphically displayed as a three-dimensional topography like a contour map as illustrated in Fig. 11-8. The near-field analysis also yields the magnitude of the effective projected solid-angle field of view Ω_{eff} in units of steradians, and the equivalent ideal half-angle field of view θ in degrees, for a circularly symmetric system.

The near-field analysis also provides for a qualitative evaluation of symmetry, smoothness, and idealness of the system response. Finally, the near-field analysis provides the function $R(\theta, \phi)$ which can be used to correct the measured flux Φ_m for the spatial distribution of $\Phi(\theta, \phi)$ to yield to the total flux Φ_T.

The far field is defined as the linearized and normalized sensor response to a point source at angles *far* from the optical axis as evaluated to many orders below the on-axis response. The extent to which this analysis must be conducted depends on the nature of the off-axis background conditions for which the sensor is being calibrated.

It is sometimes necessary to baffle the sensor response to off-axis sources to less than 10^{-10} of the on-axis response. However, the experimental evaluation of well-baffled systems is extremely difficult, as previously outlined. For this reason it is usually not practical to attempt a mapping of the response to a point source over a two-dimensional grid as in the near-field analysis. Often a single scan is sufficient. The data is graphically displayed as a normalized cross-sectional field of view as illustrated in Fig. 11-9.

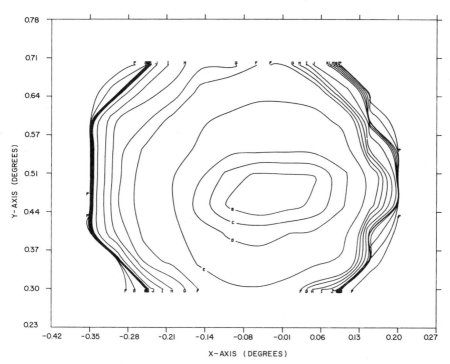

Figure 11-8 Graphical display of the near-field relative response as a three-dimensional topography like a contour map. (Intervals are $A = 1.0$, $B = 0.8$, $C = 0.6$, etc.)

The field-of-view data, for both the near field and the far field, must be linearized as previously indicated. This is accomplished by making use of the transfer function equation obtained in the "linearity analysis." As in the linearity analysis it is necessary to refer the field-of-view data to the high-gain channel by making use of the channel gain differences. It is also essential that the offset error obtained in the "dark-noise analysis" be subtracted from all measurements before the outputs are projected onto the high-gain channel, otherwise the offset error will be amplified by the gain difference.

The transfer function relates sensor output Γ to source area A (or source flux Φ which is proportional to area). The field-of-view data is linearized by substituting the measured sensor output Γ into the transfer function equation

$$A(\theta, \phi) = a\Gamma + b\Gamma^2. \tag{11-3}$$

The normalized spatial responsivity function is given by

$$R(\theta, \phi) = A(\theta, \phi)/A_0, \tag{11-4}$$

where A_0 is the peak or on-axis value of the source area.

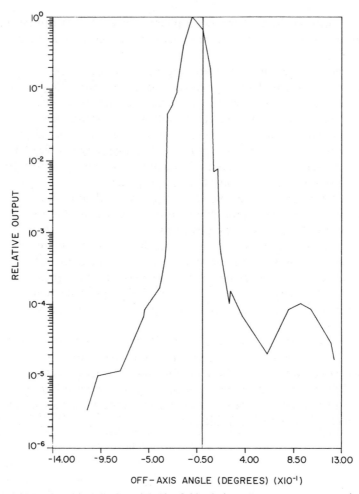

Figure 11-9 Graphical display of the far-field relative response as a cross section.

In general, the field of view is not circularly symmetric, and must be evaluated as a three-dimensional topography. This is accomplished by measuring the sensor response to a point source using a physical setup like that illustrated in Fig. 11-1. There are several alternative methods by which the spatial response can be measured and evaluated. The problem is visualized in terms of the definition of solid angle and projected solid angle as follows: The Z-axis of a unit hemisphere is visualized as lying along the axis of the collimated beam with the origin of coordinates at the sensor entrance aperture. An incremental solid angle is defined as the incremental area on the surface of the sphere but the projected solid angle is the projection of that area onto the base of the hemisphere. Thus the data can be properly represented as a three-dimensional topography like a contour map on a plane surface.

The most direct measurement approach utilizes a polar coordinate scheme. In this method the optical axis of the sensor is aligned with the collimator and the fiducial set to read zero. The sensor response is measured as the sensor is rotated in azimuthal angle which corresponds to the radius vector in polar coordinates. Then, the sensor is incremented in rotation about its own axis, which corresponds to the polar angle θ. For each increment $\Delta\theta$ the sensor is again rotated in azimuthal angle. The resulting data set can be plotted as a grid on polar graph paper.

The polar coordinate scheme suffers from several limitations. It is inconvenient and sometimes impossible to rotate the sensor about its own axis, and the grid system is expanding with radius.

A more practical measurement approach utilizes a rectangular coordinate scheme. In this method the sensor is incremented by rotation in the vertical angle, which corresponds to the Y-axis, and the sensor response is measured as the sensor is rotated in azimuthal angle, which corresponds to the X-axis. The resulting data set can be plotted on a uniform, rectangular, coordinate grid.

The rectangular coordinate scheme suffers from distortion of the rectangular grid for large off-axis angles. However, the error in calculating the projected solid angle is less than 1% for off-axis angles less than 10°.

The data, for the rectangular scheme, may be arranged into matrix form as given in Eq. (11-5), where there are n scans in the X-axis angle (azimuthal angle), each containing m sample points.

	x_1	x_2	x_3	$\cdots$	x_m
Scan 1 (y_1)	$R_{1,1}$	$R_{1,2}$	$R_{1,3}$	$\cdots$	$R_{1,j}$
Scan 2 (y_2)	$R_{2,1}$	$R_{2,2}$	$R_{2,3}$	$\cdots$	$R_{2,j}$
Scan 3 (y_3)	$R_{3,1}$	$R_{3,2}$	$R_{3,3}$	$\cdots$	$R_{3,j}$
$\vdots$	$\vdots$	$\vdots$	$\vdots$		$\vdots$
Scan n (y_n)	$R_{i,1}$	$R_{i,2}$	$R_{i,3}$	$\cdots$	$R_{n,m}$

$$(11\text{-}5)$$

The jth column refers to the jth X-axis angle and the ith row to the ith scan at the ith Y-axis angle.

The value of the incremental solid angle for each point on the grid is given by

$$\Delta\Omega_{i,j} = \sin(x_i - x_{i-1}) \sin(y_j - y_{j-1}) = \sin^2 \Delta\Psi, \qquad (11\text{-}6)$$

where for a uniform rectangular grid $x_i - x_{i-1} = y_j - y_{j-1} = \Delta\Psi$.

The total effective solid angle is given by summing all the increments over the entire region, each of which are weighted by the respective value of $R_{i,j}$. Thus

$$\Omega_{\text{eff}} = \sin^2 \Delta\Psi \sum_i^n \sum_j^m R_{i,j} \quad [\text{sr}]. \tag{11-7}$$

The solid angle Ω, based on a circularly symmetric field of view, is given by

$$\Omega = \pi \sin^2 \Theta \quad [\text{sr}], \tag{11-8}$$

where Θ is the half-angle edge of an ideal field of view. The half-angle Θ for an equivalent ideal field of view can be obtained by solving for Θ between Eqs. (11-7) and (11-8) to yield

$$\Theta_{\text{eff}} = \arcsin\left[1/\pi \sin^2 \Delta\Psi \sum_i^n \sum_j^m R_{i,j} \right]^{1/2} \quad [\text{deg}]. \tag{11-9}$$

REFERENCES

1   J. C. Kemp, A specular chamber for off-axis response evaluation of high-rejection optical baffling systems. Ph.D. Thesis, Utah State Univ., Logan, 1976.
2   R. P. Heinisch and T. S. Chou, Numerical experiments in modeling diffraction phenomena. *Appl. Opt.* **10**, 2248–2251 (1971).
3   R. P. Heinisch and C. L. Jolliffe, Light baffle attenuation measurements in the visible. *Appl. Opt.* **10**, 2016–2020 (1971).

CHAPTER

XII

Spectral Purity

12-1 INTRODUCTION

The sensor aperture is bombarded with unwanted flux from sources outside the sensor spectral bandpass. The sensor output for a spectrally pure measurement is a function of the flux originating from within the sensor spectral bandpass, and is completely independent of any flux originating from outside the region. Thus the calibration of the spectral response (or the spectral bandpass) of a sensor is considered, in this book, as a problem of spectral purity.

All practical sensors exhibit some out-of-band response, and large measurement errors can occur. The out-of-band response is a problem of filter blocking in radiometers and "order sorting" in spectrometers that utilize the interference phenomenon. The degree to which the out-of-band response must be characterized depends on the nature of the target and the background in which it is immersed. However, the peculiarities of the spectrum of the blackbody source, which is often used as a calibration standard, often magnify the problem of long wavelength leakage. Thus spectral purity is a serious calibration problem. The evaluation of the out-of-band response to many orders of magnitude below the in-band response is a difficult aspect of sensor calibration.

The "solar-blind" multiplier–phototube is an example of an ultraviolet sensor upon which visible light has little or no effect. Spectral purity problems are minimized when photon detectors are used near the maximum wavelength of their response function. This is because the response drops rapidly to zero for wavelengths just beyond the peak wavelength. However, long wavelength leakage becomes a serious problem whenever photon detectors must be used at wavelengths significantly less than the peak wavelength.

The purpose of this chapter is to deal with spectral purity. These are problems arising from instrument response that exhibit nonideal spectral responsivity, and source spectral distribution that is nonuniform (and often discontinuous) or unknown.

Often the spectral distribution of a field source of radiant flux is beyond the control of the observer, and in fact, may even be unknown. However, the instrument spectral response is somewhat amenable to design. The emphasis of this chapter is therefore placed on the definition of the spectral bandpass (both ideal and nonideal), the errors associated with the measurement of nonuniform sources, and the incomplete characterization of the sensor out-of-band response. The details of the spectral bandpass calibration, technique, equipment, and data processing are given in Chapter XIII.

12-2   SPECTRAL RESPONSE

The problem of the calibration of an electrooptical sensor with respect to spectral parameters is best understood in terms of a measurement goal. The goal of measurement of a remote radiant source with respect to spectral parameters is to obtain a measure of the total flux Φ_T in a specified (usually narrow) region about the center wavelength λ_0. The source spectral distribution of flux as a function of wavelength λ is given by $\Phi(\lambda)$ over all λ. Then the total flux is given by the integral of $\Phi(\lambda)$ over the spectral bandpass $\lambda_2 - \lambda_1$ of the sensor:

$$\Phi_T = \int_{\lambda_1}^{\lambda_2} \Phi(\lambda)\, d\lambda \quad [\Phi]. \tag{12-1}$$

The narrow region about the center wavelength λ_0 is the spectral bandpass of a radiometer (also referred to as the "half-power bandwidth" HPBW) or the *instaneous* spectral bandpass of a spectrometer at any specific wavelength λ_0 within the free spectral range (also referred to as the "instaneous half-power bandwidth" IHPBW), and has the units of micrometers (μm) (see Fig. 12-1).

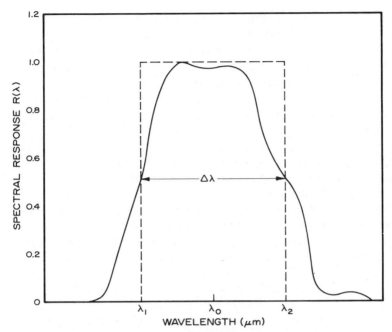

Figure 12-1 An illustration of the ideal (dashed lines) and practical (solid curve) spectral bandpass function. [Adapted from *Proc. Soc. Photo-Opt. Instrum. Eng.—Infrared Technol. II* **95**, 221 (1976).]

The actual sensor output Γ in response to $\Phi(\lambda)$ is given by the integral of the product of $\Phi(\lambda)$ with the sensor responsivity $R_0 R(\lambda)$:

$$\Gamma = R_0 \int_0^\infty R(\lambda)\Phi(\lambda) \, d\lambda \quad \text{[output]}, \tag{12-2}$$

where the integration is carried out over all wavelengths, and R_0 the peak responsivity at λ_0 (with $R(\theta, \phi)$, $R(P)$, and $R(t)$ constant).

From this, the measured target flux is obtained by the relation

$$\Phi_m = \Gamma/R_0 = \int_0^\infty R(\lambda)\Phi(\lambda) \, d\lambda \quad [\Phi], \tag{12-3}$$

which, based on Eq. (12-1), is equal to Φ_T only if

$$\int_0^\infty R(\lambda)\Phi(\lambda) \, d\lambda = \int_{\lambda_1}^{\lambda_2} \Phi(\lambda) \, d\lambda \quad [\Phi]. \tag{12-4}$$

Unfortunately, the equality in Eq. (12-4) is valid under only two special conditions, neither of which is perfectly realized in practice.

(1)   The equality in Eq. (12-4) is valid for an ideal instrument relative spectral response function $R(\lambda)$, which is defined as one that has unity response over the region $\lambda_2 - \lambda_1$ (the spectral bandpass), and is zero elsewhere. In this case, since $R(\lambda)$ is a unity constant which passes through the integral, Eq. (12-4) is identically true.

Such an ideal response function will always yield the total flux Φ_T in the bandpass regardless of the shape of $\Phi(\lambda)$, and, consequently, great effort is expended to achieve practical systems that approach the ideal. Much of the effort of both calibration and data reduction consists of qualifying the spectral characteristics of the instrument and the target.

(2)   The equality in Eq. (12-4) is valid for the case where $\Phi(\lambda) = \Phi_0$ (a constant) which requires that it be either a uniform source or a monochromatic source (approaching zero bandwidth).

Case (1), the ideal sensor, is not achieved in practical instruments, and monochromatic sources do not exist; therefore, the meaning of a measurement is given in terms of the uniform source of case (2) as follows: "Provided that $\Phi(\lambda)$ is spectrally uniform, it has a magnitude of Φ_T as calculated by Eq. (12-3)." Recommended practice is to calculate the flux from Eq. (12-3) and describe it as the *peak normalized flux* [1, 2].

Under these practical conditions [case (2)], Eq. (12-4) becomes

$$\lambda_2 - \lambda_1 = \int_0^\infty R(\lambda)\, d\lambda = \int_{\lambda_1}^{\lambda_2} d\lambda \quad [\mu m]. \tag{12-5}$$

The right-hand side of Eq. (12-5) is an expression of the "ideal" spectral bandpass where $R(\lambda)$ is unity and the limits on the integral define the edges of the band. The left side of Eq. (12-5) is an expression of the nonideal spectral bandpass.

12-3   THE IDEAL SPECTRAL BANDPASS

Equation (12-5) is the basis for a definition of both the ideal and nonideal spectral bandpass. The ideal spectral bandpass is given by

$$\lambda_2 - \lambda_1 = \int_{\lambda_1}^{\lambda_2} d\lambda \quad [\mu m], \tag{12-6}$$

which is a square function, as illustrated in Fig. 12-1 for which $R(\lambda)$ is unity between λ_1 and λ_2 and is zero elsewhere. Thus the ideal bandpass is defined either by the increment $\Delta\lambda$ or by the wavelengths λ_2 and λ_1 that correspond to the edges of the ideal square response function.

12-4   THE NONIDEAL SPECTRAL BANDPASS

The nonideal spectral bandpass is also defined by Eq. (12-5) as

$$\lambda_2 - \lambda_1 = \int_{-\infty}^{\infty} R(\lambda)\, d\lambda \quad [\mu m]. \tag{12-7}$$

A practical spectral bandpass function $R(\lambda)$ is nonideal. Figure 12-1 illustrates a peak-normalized practical spectral bandpass curve. Equation (12-7) yields the area under the $R(\lambda)$ curve, which for normalized functions is exactly equal to the width of the *equivalent* ideal square curve. Thus, for practical systems, the bandpass is defined in terms of the wavelengths λ_1 and λ_2 that give the edges of the equivalent (area) ideal square response function.

For some systems, the response function $R(\lambda)$ is very nearly *Gaussian* or *natural*. The Gaussian function is given by

$$R(\lambda) = \exp[-(\lambda - \lambda_0)^2/1.288\sigma^2], \tag{12-8}$$

where $R(\lambda)$ is unity for $\lambda = \lambda_0$ and zero for $\lambda = \infty$ and σ the half-width corresponding to $R(\lambda) = 0.460$.

The integral of $R(\lambda)$ is equal to the half-width of the equivalent ideal response function, that is,

$$\int_{0}^{\infty} \exp[-(\lambda - \lambda_0)^2/1.288\sigma^2]\, dx = \sigma \tag{12-9}$$

to within 1%. Thus the edge of the ideal equivalent response function passes through the 0.460 point on the natural function. Common practice is to specify the half-power point $R(\lambda) = 0.50$ as the equivalent ideal width as a first-order approximation for functions that are well behaved, i.e., that are Gaussian.

The practical response is said to be equivalent to the ideal response because there is equal area under each curve. Either the practical response curve or its ideal equivalent will yield the same sensor output from a source of uniform spectral distribution. This is because the response to this uniform continuum outside the limits λ_1 and λ_2 exactly compensates for the reduced response at wavelengths other than λ_0, between λ_1 and λ_2.

Most practical systems are neither ideal square functions nor Gaussian functions. It is therefore often necessary to perform an integration of Eq. (12-7) in order to accurately specify the equivalent ideal bandpass. The measured spectral bandpass function $R(\lambda)$ for most practical systems is not analytic; therefore, it is necessary to increment it as illustrated in Fig. 12-2.

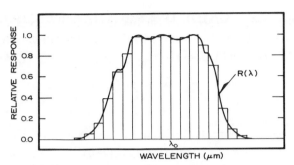

Figure 12-2   Incremented spectral bandpass function $R(\lambda)$.

Equation (12-7) is approximated as

$$\lambda_2 - \lambda_1 = \Delta\lambda \sum_{i=1}^{n} R(\lambda) \quad [\mu m], \tag{12-10}$$

where $\Delta\lambda$ is the width of each increment (a constant) and $R(\lambda)$ the corresponding midpoint value of $R(\lambda)$ for each increment.

The response function $R(\lambda)$ should be measured as a *system* function [1, p. 20]. This will include the effect of all optical components (windows, mirrors, filters) of the system including the detector. This is especially important where multiple reflections can occur between elements in an optical system.

12-5   ERRORS ASSOCIATED WITH NONIDEAL SPECTRAL BANDPASS

The conditions previously given for which the exact value of source flux can be obtained from a field measurement are

(1)   The instrument spectral bandpass is ideal.
(2)   The source is spectrally uniform.
(3)   The source is purely monochromatic.

Generally none of these conditions are achieved in any measurement; consequently, corrections must be made for the shape of the spectral distribution of the source. Large errors can occur in the presence of high intensity, out-of-band sources.

The spectral distribution of the flux can be expressed as

$$\Phi(\lambda) = k\Phi_r(\lambda), \tag{12-11}$$

where $\Phi_r(\lambda)$ is the relative spectrum (or the shape of the spectrum) and the coefficient k gives the absolute units. The sensor output can be expressed in terms of Eqs. (12-2) and (12-11) as

$$\Gamma = R_0 k \int_0^\infty \Phi_r(\lambda) R(\lambda) \, d\lambda, \qquad (12\text{-}12)$$

from which k is obtained as

$$k = \Gamma \Big/ \left(R_0 \int_0^\infty \Phi_r(\lambda) R(\lambda) \, d\lambda \right). \qquad (12\text{-}13)$$

The total flux is then obtained as

$$\Phi_T = k \int_{\lambda_1}^{\lambda_2} \Phi_r(\lambda) \, d\lambda, \qquad (12\text{-}14)$$

where λ_1 and λ_2 define any wavelength internal that satisfies the measurement objectives and for which the relative spectrum $\Phi_r(\lambda)$ of the source is available [3].

The solution to Eqs. (12-13) and (12-14) requires that $\Phi_r(\lambda)$ be known either from measurement or from theory. It is also necessary to know $R(\lambda)$, which is an important and difficult aspect of sensor calibration. If such a correction is not made, the data should be reported as *peak normalized*. In any case it is important to report the function $R(\lambda)$ so that the reader can apply his own corrections [1, p. 17].

REFERENCES

1   F. E. Nicodemus and G. J. Zissis, "Report of BAMIRAC—Methods of Radiometric Cali-
    bration," ARPA Contract No. SD-91, Rep. No. 4613-20-R (DDC No. AD-289, 375), p. 15.
    Univ. of Michigan, Infrared Lab., Ann Arbor, Michigan (1962).
2   F. E. Nicodemus, Normalization in radiometry. *Appl. Opt.* **12**, 2960–2973 (1973).
3   G. A. Ware, Atmospheric hydroxyl ($\Delta v = 2$) rotational temperatures. Ph.D. Dissertation,
    Utah State Univ., Logan, 1978.

CHAPTER

XIII

Spectral Calibration

13-1 INTRODUCTION

This chapter deals with the experimental characterization or calibration of the spectral aspects of a remote sensor of radiant flux. The problem of estimating the precision or repeatability of such measurements is also considered.

13-2 THE RELATIVE SPECTROMETER CALIBRATION

The topics covered in this section apply only to the calibration of a scanning spectrometer, including either the simple, sequential, scanning spectrometer or the multiplex spectrometer.

13-2-1 Spectral Scan Position Calibration

The objective of the spectral scan position calibration is to provide a functional relationship between the wavelength (or wave number) and the spectral scan position of the spectrometer.

For a multiplex spectrometer, all wavelengths within the free spectral range are sampled simultaneously, so it is an "effective scan position" which must be determined.

The ideal standard for this calibration, a tunable monochromatic source (the equivalent of an electronic sine wave generator), does not exist. However, this calibration can be accomplished by stimulating the sensor with an external source which has spectral features of known wavelengths. The source could consist of one or more of the following:

(1) a broadband radiant emitter with one or more narrowband interference filters,

(2) a broadband radiant emitter and a calibrated monochromator,

(3) a broadband radiant emitter and an absorption filter with lines of known wavelengths, and

(4) an emission source with a discrete line structure of known wavelengths.

A dual-channel recorder is used to simultaneously record the scanning position reference output and the primary data output which contains the instrument response to a known spectral source.

The broadband source and calibrated monochromator provide a convenient method. The resultant output signal consists of a single spectral "line." Noise may cause an apparent shift in the position of the line from scan to scan. It is therefore appropriate to find the average position of the line over a number of scans. This procedure is repeated at a sufficient number of wavelengths to determine the function

$$\lambda = f(\delta), \tag{13-1}$$

where δ is the scan position (or percentage of scan). This function can be determined through the use of standard curve-fitting techniques.

It is appropriate to estimate the precision and accuracy by noting the spread in the values and the quality of the fit. It is also necessary to establish the limits over which Eq. (13-1) is valid, i.e., the free spectral range of the spectrometer.

The spectral scan position calibration of a circular-variable filter (CVF) spectrometer is illustrated in Fig. 13-1 where the absorption lines of polystyrene plastic are observed in the spectrum of a blackbody source. The wavelength and percentage of scan of each feature constitute a data set from which Eq. (13-1) can be derived.

13-2-2 The Spectral Resolution Calibration

The objective of the spectral resolution calibration is to determine the resolving power of the spectrometer

$$\mathscr{R} = \lambda/\Delta\lambda \qquad \text{or} \qquad \bar{\nu}/\Delta\bar{\nu} \tag{13-2}$$

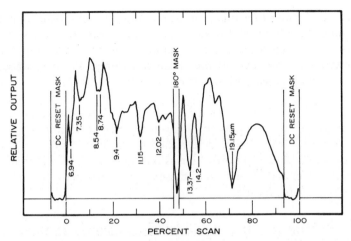

Figure 13-1 The absorption spectrum of polystyrene plastic observed with a circular variable filter spectrometer (CVF). [Adapted from C. L. Wyatt, *Appl. Opt.* **14**, 3091 (1975).]

where $\Delta\lambda$ or $\Delta\bar{\nu}$ is the width of a resolution element. The resolution element is specified as the width of the equivalent ideal instantaneous bandwidth obtained by integrating the instantaneous relative response function $R_i(\lambda)$ over all wavelengths, namely,

$$\Delta\lambda = \int_0^\infty R_i(\lambda)\,d\lambda \quad [\mu m] \tag{13-3}$$

which, for well-behaved systems, is approximately the instantaneous half-power bandwidth (IHPBW).

In general, for a spectrometer with n resolution elements it will be necessary to characterize $R_i(\lambda)$ at n separate wavelengths, although for many spectrometers the resolution is constant when calculated in the appropriate units. For example, the resolution $\Delta\bar{\nu}$ is constant throughout the free spectral range of an interferometer–spectrometer. However, the resolution varies continuously with wavelength or wave number for a circular-variable filter spectrometer.

Both the central wavelength λ_0 and the shape of the bandpass $R_i(\lambda)$ may be functions of the geometrical properties of the source, the illumination of the instrument collector, and the monochromator temperature. The general rule, that spectrometers should be calibrated under conditions that approximate, as nearly as possible, the conditions under which field measurements are to be made, applies here.

The instantaneous spectral bandpass function $R_i(\lambda)$ is obtained as a normalized functional relationship between the sensor output Γ and the wavelength λ. A convenient experimental method to obtain this relationship is to stimulate the spectrometer with a broadband radiant emitter coupled to a calibrated monochromator. The monochromator can be set at any wavelength λ_0 and the spectrometer scanned through its free spectral range. The spectrometer output as a function of wavelength is the function $R_i(\lambda)$ at the wavelength λ_0 provided that the width of the monochromator "line" is small compared to the spectrometer resolution. This process can be repeated at any number of wavelengths throughout the free spectral range.

The resolution can also be determined by stimulating the spectrometer with a spectral source of known "emission lines," such as a mercury lamp. The identification and resolution of these known lines provide a measure of the spectrometer resolution.

13-2-3   The Relative Spectral Responsivity Calibration

The responsivity of a spectroradiometer is given by

$$R(\lambda) = R_0 R_s(\lambda) R(\theta, \phi) R_i(\lambda) R(t) R(P), \qquad (13\text{-}4)$$

where R_0 is the absolute responsivity in units of sensor output per unit flux, $R_s(\lambda)$ the instrument function (relative to the free spectral range), $R(\theta, \phi)$ the relative spatial responsivity (the field of view), $R_i(\lambda)$ the instantaneous relative spectral responsivity, $R(t)$ the relative temporal responsivity, and $R(P)$ the relative polarization responsivity. The value of $R_s(\lambda)$ varies with wavelength because the detector responsivity, the transmittance and reflectance of mirrors, lenses, gratings, etc., all vary with wavelength.

The radiometer is considered a degenerate case of the spectroradiometer fixed at a single wavelength. In this case $R_s(\lambda) = 1$ and Eq. (13-4) becomes

$$R = R_0 R(\theta, \phi) R_i(\lambda) R(t) R(P). \qquad (13\text{-}5)$$

The objective of the relative responsivity calibration of a high-resolution spectrometer is to determine the function $R_s(\lambda)$. The detailed evaluation of $R_s(\lambda)$ is considered with that of R_0 in Section 13-7.

13-3   BANDPASS CALIBRATION OF A RADIOMETER

For relatively wide-band radiometers, knowledge of the shape of $R_i(\lambda)$ is essential if corrections are to be made for the shape of $\Phi(\lambda)$ as outlined in Chapter XII.

The instantaneous spectral bandpass function $R_i(\lambda)$ cannot be calculated as the product of the optical transmittance (or reflectance) of the filters, lenses, mirrors, etc., of the system. This is because multiple reflections alter the spectral characteristics of these components. Therefore, this function must be measured as a system.

The instantaneous spectral bandpass function $R_i(\lambda)$ is obtained as a normalized functional relationship between the output signal Γ and wavelength λ. A convenient method to obtain this relationship is to stimulate the radiometer with a broadband radiant emitter and a calibrated monochromator. The monochromator is "tuned" through the radiometer spectral bandpass and the output recorded and correlated with wavelength.

The difficulty with this procedure is that the monochromator radiant flux is not a constant output of wavelength. The shape of the monochromator spectral intensity will weight the shape of the radiometer and must be corrected. This is usually accomplished with a well-qualified standard detector [1, 2].

13-4   THE CALCULATION OF THE NORMALIZED FLUX

Standard sources of radiant flux for calibration standards emit energy continuously at all wavelengths in accordance with Planck's equation. Not all the energy emitted by the standard source is equally effective in producing an output response in the sensor. The value of $\Phi(\lambda)$ is weighted by the shape of the instantaneous spectral bandpass function $R_i(\lambda)$ of the sensor at any wavelength λ within the free spectral range of a spectrometer, or of the spectral bandpass of a radiometer.

The weighted value of the spectral flux $\Phi_N(\lambda)$ is given by

$$\Phi_N(\lambda) = (\lambda_2 - \lambda_1)^{-1} \int_0^\infty R_i(\lambda)\Phi(\lambda) \, d\lambda \quad [\Phi \; \mathrm{m}^{-2} \; \mathrm{sr}^{-1} \; \mu\mathrm{m}^{-1}] \quad (13\text{-}6)$$

for a spectrometer, where $\lambda_2 - \lambda_1$ is the equivalent ideal instantaneous spectral bandwidth (IHPBW), $R_i(\lambda)$ the instantaneous spectral bandpass function (with the spectrometer stopped at λ), and $\Phi(\lambda)$ given by Planck's equation.

Likewise, the value of the band flux Φ_N is given by

$$\Phi_N = \int_0^\infty R_i(\lambda)\Phi(\lambda) \, d\lambda \quad [\Phi \; \mathrm{m}^{-2} \; \mathrm{sr}^{-1}] \quad (13\text{-}7)$$

for a radiometer.

Equations (13-6) and (13-7) provide a measure of how effectively each increment of $\Phi(\lambda)$ can stimulate a response in the sensor. This, of course, depends on $R_i(\lambda)$. Hence, the value of $\Phi_N(\lambda)$ or Φ_N will depend on how $R_i(\lambda)$

was normalized. Ordinarily, for narrow-band spectrometers and radiometers, $R_i(\lambda)$ is *peak* normalized. In this case the weighted flux is termed *peak normalized flux*. However, in some cases, particularly for wide-band radiometers that contain some irregularities in the shape of the bandpass function $R_i(\lambda)$, it may be appropriate to normalize to some other level such as the average between λ_1 and λ_2. Such a wide-band filter is illustrated in Fig. 13-2 where peak normalization would appear to be most inappropriate.

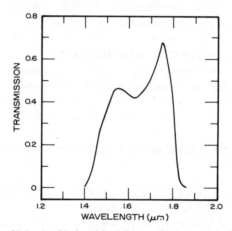

Figure 13-2   A wide-band interference filter transmission curve.

The instantaneous half-power bandwidth (IHPBW) is defined in terms of the equivalent ideal bandpass which is obtained by

$$\lambda_2 - \lambda_1 = \int_0^\infty R_i(\lambda)\, d\lambda \quad [\mu m]. \tag{13-8}$$

For high-resolution spectrometers the IHPBW tends to be small compared to the variations in the blackbody source $\Phi(\lambda)$ so that $\Phi(\lambda) = \Phi_0$, a constant; then

$$(\lambda_2 - \lambda_1)^{-1} \int_0^\infty R_i(\lambda)\, d\lambda = 1 \tag{13-9}$$

in Eq. (13-6) so that

$$\Phi_N(\lambda) = \Phi(\lambda). \tag{13-10}$$

The flux Φ in the preceding analysis represents, as appropriate, the sterance [radiance, luminance] or the areance [irradiance, illuminance].

Spectral purity can be a serious problem when calculating the effective flux if $R_i(\lambda)$ has not been well characterized outside the nominal IHPBW.

13-5 SPECTRAL PURITY

The determination of the normalized flux within the IHPBW of a spectrometer or radiometer depends on the instantaneous relative spectral function $R_i(\lambda)$. The determination of $R_i(\lambda)$ is one of the more difficult aspects of the calibration of optical spectrometers and radiometers, because all practical systems exhibit some degree of out-of-band response which must be accounted for.

13-5-1 Introduction

Problems of spectral purity arise in a calibration exercise in part from the unique properties of blackbody radiation. This is illustrated in the case of the calibration of a long-wave infrared circular-variable filter spectrometer.

In keeping with the philosophy that spectrometers should be calibrated under conditions that approximate, as nearly as possible, the conditions under which field measurements are to be made, it is desirable to provide an extended area source for instruments designed to measure an extended source, and, conversely, to provide a point source calibration for instruments designed to measure a point source.

Practical IR instruments are frequently so sensitive that an ambient temperature (or greater) extended source may saturate the instrument. This is especially true of cryogenic IR systems where special techniques must be used [3].

Figure 13-3 illustrates the response of a cryogenic CVF spectrometer to an extended area cold blackbody source. The instrument response il-

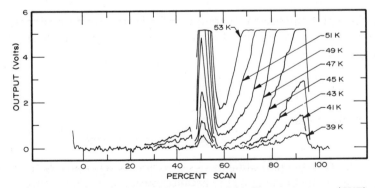

Figure 13-3 The response of an LWIR circular variable filter spectrometer (CVF) over the range 7–24 µm to an extended-area blackbody source. [Adapted from C. L. Wyatt, *Appl. Opt.* **14**, 3090 (1975).]

lustrated in Fig. 13-3 records a significant response in the 20–22 μm range (90–100% scan position) to a 39K blackbody source, and is indicative of the system sensitivity. There is evidence of a serious spectral leak in the 12.0–13.0 μm region (50–55% scan position). In addition to leakage problems, the extended area source provides a calibration only in the long-wavelength part of the free spectral range because of the long-wavelength shift of the emitted energy in a low-temperature blackbody.

The wavelength for which maximum blackbody radiation occurs is given by

$$\lambda_M = 2898/T \quad [\mu m]. \qquad (13\text{-}11)$$

For calibration purposes, λ_M is significant. At wavelengths less than λ_M the following conditions apply.

(1)   The power spectral density is changing very rapidly with λ, so that a slight error in determining the temperature results in a relatively large error in determining the radiation.

(2)   The power spectral density is changing very rapidly with λ, so that (as illustrated in Fig. 13-3) the power spectral density almost has the appearance of a step function. Thus slight errors in determining the spectral scan position can result in relatively large errors in the magnitude of the output deflection.

The alternative is to make use of a point source at a distance, which permits the use of a higher temperature blackbody, provided that the incident flux density, at the instrument aperture, can be sufficiently reduced by the effect of $1/s^2$ and reduced source area. Then, when λ_M is less than any wavelength in the free spectral range, the slope of $\Phi(\lambda)$ as a function of λ is relatively flat. Figure 13-4 shows the response of the CVF spectrometer to a

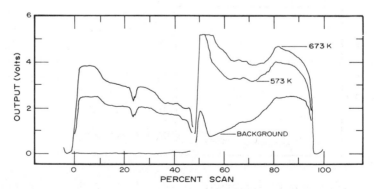

Figure 13-4   The response of a LWIR circular variable filter spectrometer (CVF) over the range 7–24 μm to a distant small-area blackbody source. [Adapted from C. L. Wyatt, *Appl. Opt.* **14**, 3090 (1975).]

point source at 573 and 673K. The background reading shown in Fig. 13-4 was obtained with the blackbody aperture closed and represents the spectral flux from the cold aperture (about 60K) which must be subtracted from the blackbody data. The use of a small aperture at a distance attenuates the incident flux, and provides nonsaturated response to the higher temperature blackbody radiation. It is noted that the 13-μm leak is still very evident in these data. The 573 and 673K spectral power curve peaks at wavelengths less than 5 μm; consequently, the curve is relatively flat resulting in a useful output throughout the entire free spectral range of the CVF spectrometer.

The point source calibration can be extrapolated to obtain an extended source calibration through the use of the invariance of sterance [radiance, luminance].

The 12–13-μm leak, evident in the extended area blackbody calibration of Fig. 13-3, is analyzed as follows: When the CVF is stopped at the 12.5-μm position, the instantaneous $R_i(\lambda)$ function has a leak response at some other wavelength. It is suspected that the filter is leaking at 26 μm since the detector is responsive there and since interference filters are difficult to block at wavelengths at and beyond 2.0 times the bandpass wavelength. Inspection of Fig. 13-3 reveals that the response to the 39K extended area blackbody at 12.5 μm (50% scan position) is about equal to the response at 22 μm (86% scan position). However, the spectral sterance [radiance] of the blackbody at 22 μm is 2×10^4 times as great as at 12.5 μm (all other things being equal). The normalized flux calculated by Eqs. (13-6) and (13-7) would be in error by many orders of magnitude using this extended area source data at 12.5 μm unless $R_i(\lambda)$ was characterized for an out-of-band response.

In this particular example it is noted that the ratio of the spectral sterance of the 39K blackbody at 26 μm (the possible leak wavelength) to that at 12.5 μm is 1×10^5. Thus, in this case of a 39K blackbody calibration, there would still be an error of a factor of 2 if the magnitude of the spectral response function $R_i(\lambda)$, in the leak region (26 μm), was 10^{-5} relative to the bandpass at 12.5 μm, provided the width of the leak is the same as the bandpass at 12.5 μm. Usually the leak is quite broad. Thus, to use the 39K data to calibrate the LWIR CVF at 12.5 μm requires that $R_i(\lambda)$ be known to 10^{-6} or 10^{-7} relative to the peak response. Unfortunately the present state-of-the-art limits direct measurements of $R_i(\lambda)$ to about 10^{-4}.

13-5-2   Out-of-Band Leakage

The out-of-band leakage, illustrated in the previous section, is typical of higher order resonance modes in a thin-film interference filter. Ordinarily, higher order blocking is achieved by depositing the interference layer upon substrate materials that provide *absorption* blocking.

There exists in nature a great many optical materials that provide adequate short-wavelength blocking by absorption; conversely, very few materials are available that provide adequate long-wavelength blocking. Consequently, long-wavelength blocking is a major problem in the design and utilization of interference filters.

All monochromators that utilize the phenomena of interference, such as thin-film filters, gratings, and interferometers, are subject to order problems. In general, adequate blocking by interference techniques is limited to wavelengths less than two times the bandpass wavelength. Various "order sorter" techniques are used including combining interferometers or gratings with thin-film interference filters or with absorptive optical materials.

Another case of out-of-band leakage is the phenomenon of scattering, which becomes a problem when out-of-band energy reaches the detector by scattering paths that bypass the basic monochromator device. The effects of scattering are minimized through the use of good layout design including extensive baffling.

Another cause of out-of-band leakage is the phenomenon of fluorescence in which the absorption of radiation at one wavelength results in the emission of radiation at another wavelength. This is a common problem associated with ultraviolet instruments. The high-energy ultraviolet radiation may cause lenses, gratings, prisms, or other optical devices to fluoresce in the visible.

13-5-3   Calibration Techniques

The major concern of this section is the consideration of calibration techniques that permit the detection of leakage, when it exists, and the consideration of methods of minimizing the impact of leakage on obtaining an absolute calibration.

The unique characteristics of blackbody emission may be utilized to reveal the presence of out-of-band leakage, as was illustrated in Section 13-5-1 for the CVF spectrometer. The long-wavelength shift of the wavelength of maximum radiation with reduced blackbody temperature may be utilized to test the accuracy of the shape of the relative spectral response function curve and the effects of out-of-band leakage.

During the calibration process it is necessary to calculate the normalized spectral flux. This calculation depends on the response function $R_i(\lambda)$. If there is an error in the determination of $R_i(\lambda)$ caused by shape factors in the bandpass, or out-of-band leakage, the calculated value of $\Phi_N(\lambda)$ will be in error.

The absolute responsivity R_0 which is the ratio of the sensor output Γ to the normalized flux $\Phi_N(\lambda)$ should be a constant for linear systems (when the offset error is zero). Hence, the accuracy of $R_i(\lambda)$ can be determined by calculating R_0 for a series of calibration tests in which the temperature of the blackbody is varied. As the blackbody temperature is reduced, the energy shifts to longer wavelengths. Thus if there is a long-wavelength leak, there will be a systematic increase in the responsivity that will make the instrument appear more sensitive at correspondingly lower blackbody temperatures. If there are shape errors, they may show up as random fluctuations in the value of R_0 and give another measure of the calibration error.

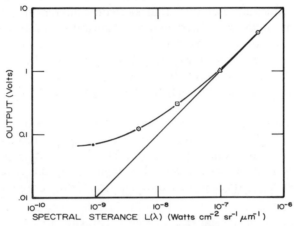

Figure 13-5 An example of long-wavelength filter leakage for various temperatures: 1200K (⊙), 1000K (△), 800K (□), 700K (⊙), and 600K (●).

Figure 13-5 illustrates long-wavelength leakage in a radiometer where the output voltage is plotted as a function of source spectral sterance [radiance] on a log–log scale. The use of the log–log scale results in a 45° slope (the responsivity R_0) for linear systems with zero offset. The apparent nonlinearity in Figure 13-5 results from long-wavelength spectral leakage. This graph shows leakage (excessive output) on 600, 700, and 800K calibrations. At higher temperatures the blackbody radiation is shifted into the bandpass relative to the leak until the leakage becomes negligible. In this case the linear line is drawn through the high-temperature points for which no leakage is evident. This linear asymptote yields the best value of R_0. However, it may be impossible to identify the position of the asymptote if the shift with temperature continues for all points. In this case the leakage is excessive and a calibration may not be possible.

Another method to detect the presence of long-wavelength leakage makes use of the fact that very good short-wavelength absorptive blocking substrate materials are available. These substrates, combined with interference coatings, yield long-pass filters at almost any cutoff wavelength desired. Selection of such a long-pass filter that just blocks the bandpass of the instrument permits isolation of the leakage contribution to the output in a calibration setup by simply interposing the blocker in the optical path. The instrument output will fall to zero if there is no leakage.

13-6   ABSOLUTE CALIBRATION

The experimental calibration of an electrooptical sensor to determine the absolute responsivity R_0 is considered in this section.

13-6-1   Introduction

The responsivity of a spectrometer is given by

$$R(\lambda) = R_0 R_s(\lambda)R(\theta, \phi)R_i(\lambda)R(P)R(t). \qquad (13\text{-}12)$$

The ideal standard source for the determination of the absolute responsivity R_0 would be a well-qualified unpolarized point source that is monochromatic and tunable, the equivalent of an electronic sine wave generator. Such a source could be physically adjusted in space, time, polarization, and wavelength so that each relative term in Eq. (13-12) equals unity (a monochromatic source in the spectral domain is equivalent to a point source in the spatial domain). Then R_0 could be measured independently of all the relative responsivity terms in Eq. (13-12).

A monochromator and a continuous source, or a tunable laser, calibrated with a standard detector is a possibility for such experiments. However, in the majority of cases it is necessary to use a blackbody standard as the source. As a result, the relative instantaneous spectral response $R_i(\lambda)$ weights the flux in a complex way. The blackbody radiates at wavelengths far removed from the spectral bandpass including regions where the filter blocking may not be adequate; consequently spectral purity becomes an important aspect of calibration. The following subsections describe different physical arrangements whereby blackbody standard sources are used to determine the constant R_0.

13-6-2   The Extended-Area Source

A direct sterance [radiance, luminance] calibration can be achieved with an extended area blackbody source. The source must be large enough to completely illuminate the system field stop as shown in Fig. 13-6. Actually

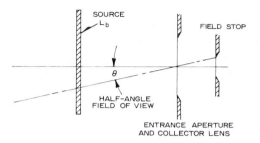

Figure 13-6   Illustration of extended-area source calibration.

the source must be large enough to completely illuminate the system throughout its entire spatially responsive domain. The inverse sterance responsivity $(1/R_0)_L$ can be computed as the ratio of the flux L_b to the output voltage V:

$$(1/R_0)_L = L_b/V \quad [\text{W m}^{-2}\text{ sr}^{-1}\text{ V}^{-1}], \qquad (13\text{-}13)$$

where L_b is the sterance [radiance] (watts per square meter steradian) of the blackbody source. This is true because of the invariance of sterance [radiance] L_b which therefore must have the same value at the entrance aperture as it does at the source. Consequently, the value of $(1/R_0)_L$ can be computed independently of the optical parameters of the instrument.

13-6-3   The Distant Small-Area Source

A direct areance [irradiance, illuminance] calibration can be achieved with a small area blackbody source. The source must be small enough so that the image of the source is completely contained within the instrument field stop as illustrated in Fig. 13-7, where I_b is the radiant pointance [intensity] (watts per steradian) of the blackbody source. The areance at the entrance aperture is given by

$$E = I_b/s^2 \quad [\text{W m}^{-2}], \qquad (13\text{-}14)$$

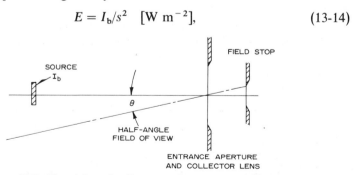

Figure 13-7   Illustration of a distant small-area source calibration.

where s is the distance between the source and the entrance aperture. The inverse areance responsivity $(1/R_0)_E$ can be computed as the ratio of the flux E to the output voltage V:

$$(1/R_0)_E = E/V   [\text{W m}^{-2} \text{ V}^{-1}], \tag{13-15}$$

and can be calculated independently of the optical properties of the instrument.

The invariance of L along a ray may be used as before to extrapolate these results to the radiance and power responsivities.

13-6-4   The Use of a Collimator

There are two problems associated with the use of the distant small area source.

(1)   It may be difficult to determine the exact location of the plane of the entrance aperture and hence the distance s [Eq. (13-14)].

(2)   Placing the source at a *distance* is not only inconvenient, in an area of limited extent, but the resulting irradiance may be insufficient for calibration purposes.

These problems are alleviated by the use of a collimator. The small area source is placed at the focus of the collimator, as shown in Fig. 13-8, where

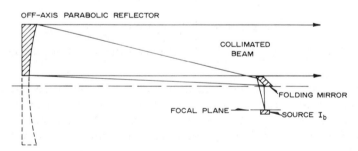

Figure 13-8   An off-axis parabolic collimator and blackbody source.

an off-axis parabolic collimator is illustrated, and the instrument is positioned within the collimated beam.

The areance within the collimated beam is given by

$$E = I/F^2   [\text{W m}^{-2}], \tag{13-16}$$

where F is the focal length of the collimator. Thus with a collimator the exact location of the instrument entrance aperture plane is not required.

However, the beam is not perfectly collimated; the divergence $\Delta\phi$ within the beam is given by

$$\text{arc } \tan(D_s/2F) = \Delta\phi \qquad (13\text{-}17)$$

where D_s is the source diameter.

13-6-5 The Near Small Area Source (Jones' Method)

The near small area source, or Jones' method, is a more complex method in that it requires some knowledge of instrument optical parameters. Nevertheless, this method may be very useful when used in some special applications. This is especially true of instruments that must be reduced in sensitivity to ambient backgrounds in order to effect a calibration.

The source must be small with respect to the area of the entrance aperture and it must be entirely within the cone formed by the entrance aperture and the point p, as shown in Fig. 13-9. Then the instrument field stop will be

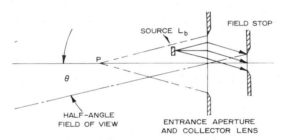

Figure 13-9 Illustration of the near small-area source (Jones' method) calibration.

completely illuminated. The instrument is focused at a distant target (infinity) as usual during this calibration so that no image of the source is formed at the field stop. The only flux arriving at the field stop from the source is that which is radiating within the instrument field of view Ω_c, and is given by

$$P = L_b A_s \Omega_c \quad [\text{W}], \qquad (13\text{-}18)$$

where L_b is the source sterance [radiance, luminance] and A_s the source area. The near small area source will produce the same output as an extended source of sterance L given by

$$L = L_b A_s \Omega_c / A_c \Omega_c = L_b A_s / A_c \quad [\text{W m}^{-2}\text{ sr}^{-1}], \qquad (13\text{-}19)$$

where the invariance of sterance is used again. Thus the effective collector area A_c must be known in this method; however, Ω_c does not enter into the calculation of L nor is it necessary to know the exact location of the entrance aperture.

13-7   SPECTRAL RESPONSIVITY CALIBRATION

The principles and experimental procedures given in the preceding sections, pertaining to the determination of the relative spectral responsivity $R_s(\lambda)$ and the absolute responsivity R_0 of a spectrometer, are illustrated in this section.

There exists such an intimate relationship between $R_s(\lambda)$ and R_0 because of the continuum nature of the standard blackbody source that they must be evaluated more or less simultaneously.

The radiometer is essentially a degenerate case of the spectrometer stopped at a fixed wavelength. In this case $R_s(\lambda)$ equals unity and the same data-processing procedures hold.

13-7-1   Weighting the Transfer Function

The starting point is the transfer function obtained in the *linearity analysis*. It is an equation of second degree that relates sensor output Γ to source area A (to avoid problems of spectral purity) as

$$A = a\Gamma + b\Gamma^2, \tag{13-20}$$

where the constant a describes the linear range of the transfer function and the constant b describes the nonlinear range. (Most detectors–preamplifiers exhibit some degree of nonlinearity in the large signal output range.)

The objective of the absolute calibration is to "anchor" Eq. (13-20) to the absolute value of the flux Φ. There is a constant $k(\lambda)$ such that

$$\Phi(\lambda) = k(\lambda)(a\Gamma + b\Gamma^2). \tag{13-21}$$

The experimental arrangement, described in the preceding section, that best approximates the measurement condition for which the sensor is being calibrated is used to obtain an absolute relationship between the incident flux and the sensor output.

A simple scan, yielding the output voltage corresponding to the incident flux for each wavelength in the free spectral range would provide enough information to calculate $k(\lambda)$ at each wavelength. However, a redundancy of data is necessary to test for spectral purity and to determine the precision (or repeatability). This is accomplished, as described in the section on spectral purity, by holding the source area constant and varying the source temperature. This yields a set of points for each wavelength given by

$$\Phi_i = (1/R_0)\Gamma_i \qquad \text{at} \quad T_i, \tag{13-22}$$

where i is the index number that runs from 1 to n (the number of points in the set) and Φ_i the normalized flux corresponding to the blackbody source temperature T_i that produces an output Γ_i.

The problem is to find the value of $k(\lambda)$ in Eq. (13-21) that provides the best fit to the data set in Eq. (13-22). This is accomplished as follows: A corresponding set of source area equations is produced by substituting the output Γ_i of Eq. (13-22) (referred to the same channel) into Eq. (13-20) to yield

$$A_i = a\Gamma_i + b\Gamma_i^2. \tag{13-23}$$

The relationship between Φ_i of Eq. (13-22) and A_i of Eq. (13-23) is linear because it is generated by the functions $A(V)$ and $\Phi(V)$ which exist on the same manifold (that is, they have been referred to the same channel); thus the constant k is found for the equation

$$\Phi = k(\lambda)A \tag{13-24}$$

that gives the best fit to the data set.

The term $k(\lambda)$ has a different value at each wavelength throughout the free-spectral range of the spectrometer; thus the inverse relative spectral responsivity $1/R_s(\lambda)$ is given by

$$C(\lambda) = 1/R_s(\lambda) = k(\lambda)/k_0 \tag{13-25}$$

where k_0 is the *minimum* value of k and $C(\lambda)$ varies from unity upward.

The resulting calibration equation is written as

$$\Phi(\lambda) = C(\lambda)(k_0 a\Gamma + k_0 b\Gamma^2) \tag{13-26}$$

where the constant $k_0 a$ is the (linear) inverse absolute responsivity coefficient

$$k_0 a = 1/R_0 \tag{13-27}$$

and $k_0 b$ the nonlinear coefficient.

13-7-2   Best-Fit Determination of $k(\lambda)$

The value of $k(\lambda)$ in Eq. (13-24) is determined by best-fit techniques. The quality of the fit is a measure of the spectral purity and repeatability. A graph of Eq. (13-24) at each wavelength in the free spectral range will yield a zero-intercept linear line only if the flux Φ_i has been properly normalized.

Excessive values of the variance at one wavelength relative to another may reveal the presence of spectral impurity or of poor precision.

The relative difference D_i of the data point Φ_i is given by

$$D_i = (\Phi_i - kA_i)/\Phi_i. \tag{13-28}$$

The relative difference multiplied by 100 yields the percent difference. The error is given by

$$\sigma^2 = \sum_{i=1}^{n} D_i^2/(n-1) \tag{13-29}$$

which is the variance—a measure of the quality of the fit of Eq. (13-24) to the data set. The number of points n in the set is a constant which passes through the summation; thus the quantity that should be minimized is

$$\text{error} = \sum_{i=1}^{n} D_i^2. \tag{13-30}$$

This is accomplished by taking the derivative of Eq. (13-30) with respect to the constant k. Setting the derivative equal to zero yields

$$k = \sum_{i=1}^{n} A_i/\Phi_i \left/ \sum_{i=1}^{n} A_i^2/\Phi_i^2 \right. . \tag{13-31}$$

REFERENCES

1 Staff Report, Detectors and radiometry. *Opt. Spectra* January, pp. 34–35 (1976).
2 J. Geist and W. R. Blevin, Chopper-stabilized null radiometer based upon an electrically calibrated pyroelectric detector. *Appl. Opt.* **12**, 2532–2535 (1973).
3 C. L. Wyatt, Infrared spectrometer: Liquid-helium-cooled rocketborne circular-variable filter. *Appl. Opt.* **14**, 3089 (1975).

CHAPTER

XIV

Temporal Response

14-1 INTRODUCTION

The ability of an electrooptical sensor to detect faint levels of incident flux depends on the time taken to make an observation. However, the source may be changing with respect to time or the instrument may be moving with respect to the source, so there usually exists a limitation to the amount of time available for a measurement.

The goal of the measurement of a remote radiant source with respect to time is to obtain a measurement of the flux at a specific time (or as a function of time) or as a function of position (in time) for moving targets or sensors. The measurement of the temporal response of a sensor is relatively simple and involves the use of modulated sources, either luminescent diodes or chopped thermal sources. The sensor response as a function of time is somewhat amenable to design; therefore the emphasis of this chapter is placed on the characteristics and evaluation of the sensor design.

14-2 TEMPORAL–FREQUENCY RESPONSE

The *integration time* of a sensor is the amount of time available to obtain a measurement during which there is essentially no change in the level of the signal. The integration time can be visualized by letting the signal to be

145

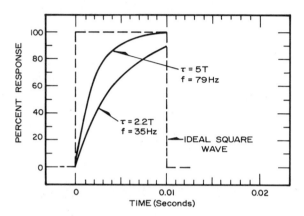

Figure 14-1   Time response to a square wave signal (of 0.01-sec duration) of a low-pass RC filter circuit where $\tau = 2.2T$ and $\tau = 5T$.

resolved be an ideal square wave pulse of width τ (as illustrated in Fig. 14-1) where τ is the integration time.

The sensor response to this square wave is an exponential given by [1]

$$100(1 - e^{\tau/T}) \quad [\%], \tag{14-1}$$

where T is the system time constant. The response to the square function may be specified in terms of the *rise time* T_r which is defined [2] as the time required for the output to change exponentially from 10 to 90% of its final value, free from any overshoot, which occurs when

$$\tau = T_r = 2.2T \quad [\text{sec}]. \tag{14-2}$$

The solution to Eq. (14-1) is as follows: When the integration time $\tau = 2.2T$, the output achieves 80% of the amplitude of the source; however, when the integration time is $5T$, the output achieves 99.3% of the source amplitude as shown in Fig. 14-1 and represents a nominal design value for accurate systems.

The electrical frequency response is related to the time constant $T = RC$ for a simple RC network by [1, p. 256]

$$f = \frac{1}{2\pi RC} = \frac{1}{2\pi T} = \frac{0.35}{T_r} \quad [\text{sec}]. \tag{14-3}$$

Figure 14-2 illustrates the frequency response for the cases where $\tau = 2.2T$ and $\tau = 5T$.

The *resolving power* of a spectrometer [3] is given by

$$\mathscr{R} = \lambda/\Delta\lambda. \tag{14-4}$$

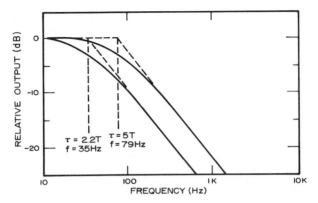

Figure 14-2 Frequency response to a constant sinusoidal wave signal of a low-pass *RC* filter circuit where $\tau = 2.2T$ and $\tau = 5T$.

The *dwell* time t_d is defined as the time that the spectrometer dwells on a single resolution element in a simple sequentially scanning spectrometer. The dwell time is given by

$$t_d = 1/\mathscr{R}S \quad [\text{sec}], \tag{14-5}$$

where S is the scan rate (scans per second). The dwell time is the integration time for a sequentially scanning spectrometer. Equations (14-2) and (14-5) may be combined for the condition $\tau = t_d = 5T$ to give a rule of thumb for f the *electrical frequency bandpass*:

$$f = 5\mathscr{R}S/2\pi \cong 0.8\mathscr{R}S \quad [\text{Hz}], \tag{14-6}$$

which will reproduce the resolution element within 99% of its true amplitude.

Good design practice dictates that the time constant be set at a value equal to or less than the dwell time divided by 5, even though this results in a loss of instrument sensitivity. This is because the time constant can *always* be increased—after the fact of the measurement—by various techniques of data processing, but it is *never* possible to decrease the time constant by post-measurement data processing.

The *electrical frequency bandpass* of the spectrometer is measured by observing the output response to a modulated light source. The two modulation techniques used in this phase of calibration are the transient response to a step function and the steady-state response to sine waves of various frequencies. Solid-state, light-emitting, diode sources and mechanical choppers are often used to accomplish the modulation.

In the transient response method, the instrument is stimulated with the output of a luminescent diode which is driven with a repetitive square wave pulse. An indium arsenide (InAs) diode, for example, emits at 3.2 μm in the IR and has a rise time of less than 100 nsec. The rise time of the observed detector response can be used to calculate the frequency response using Eq. (14-3) provided there is no overshoot present. Luminescent diodes are available in the visible, near, and intermediate IR.

Variable-speed sine wave choppers have also been used to produce a continuously variable sine wave modulated output.

The evaluation of the frequency response of a spectrometer is complicated by the dynamics of the spectral scan. However, a luminescent diode can be mounted internally adjacent to the IR detector to produce a square wave pulse response that is independent of the spectral scan. In this case the frequency response of a system can be evaluated at any convenient time by activating the internal diode source.

14-3   TEMPORAL RESPONSE—AN EXAMPLE

The operation of helium-cooled photoconductive detectors under conditions of zero effective background requires the use of special frequency-compensating preamplifiers [4], and the performance is often complicated by multiple time constant phenomena. Such detectors have been employed in cryogenic high-vacuum radiometers and spectrometers used for *in situ* measurements of upper atmospheric emissions [5].

A mercury-doped germanium detector and transimpedance amplifier (TIA) were tested for frequency response as follows: The detector–preamplifier module was mounted in a 1-liter helium-cooled Dewar. The detector aperture was equipped with an adjustable cold shield to provide various background rates including an effective zero level. The Dewar was also equipped with a CdTe window to facilitate a blackbody calibration. An indium arsenide emitter was mounted so that the detector could be irradiated with a square wave repetitive pulse. The emitter could serve both as a transfer standard for a chopped blackbody and to measure the temporal response of the detector under various background conditions [6].

The detector response to a square wave of radiant flux exhibited bias-dependent, multiple time constant response, as shown in Fig. 14-3, as follows: The fast portion of the response (preamplifier limited) exhibits a rise time of about 3×10^{-3} sec which is not bias dependent for voltages above 4 V. This corresponds to a frequency response of about 116 Hz. The so-called "dielectric relaxation" time constant is indicated by the continued

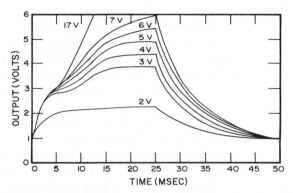

Figure 14-3   Response of a GeHg detector to a square wave radiant flux pulse illustrating multiple time constant phenomena.

but slower rise of the signal after 5.0 msec. The fact that the pulse starts at 1-V rather than zero is evidence of a third time constant characteristic of the detector [7]. Apparently carriers are trapped within the detector causing the detector resistance to assume an average level that is equivalent to a nonzero background. These trapped carriers are swept out of the detector at a rate corresponding to a time constant of approximately 85 sec. This is determined by observing the fall-time (to zero) characteristics when the emitter is turned off (not shown).

The use of a multitime constant detector in a radiometer or spectrometer could pose special problems since the system responsivity is dependent on the temporal characteristics of the source in a rather complex manner.

REFERENCES

1   R. J. Smith, "Electronics: Circuits and Devices," p. 69. Wiley, New York, 1973.
2   D. L. Metzger, "Electronic Circuit Behavior," p. 235. Prentice-Hall, Englewood Cliffs, New Jersey, 1975.
3   R. J. Bell, "Introductory Fourier Transform Spectroscopy," p. 249. Academic Press, New York, 1972.
4   C. L. Wyatt, D. J. Baker, and D. G. Frodsham, A direct coupled low noise preamplifier for cryogenically cooled photoconductive in detectors. *Infrared Phys.* **14**, 165–176 (1974).
5   C. L. Wyatt, Infrared spectrometer: Liquid-helium-cooled rocketborne circular-variable filter. *Appl. Opt.* **14**, 3089 (1975).
6   C. L. Wyatt, "Infrared Helium-Cooled Circular-Variable Spectrometer, Model HS-1," Final Rep., AFCRL-71-0340, Contract No. F19628-67-C-0322, Electro-Dynamics Lab., Utah State Univ., Logan (1971).
7   C. R. Jeffus, M. M. Blouke, E. E. Harp, and R. L. Williams, Operating characteristics of extrinsic Ge at helium temperatures. *Proc. Infrared Inform. Symp.* **15**, 323–346 (1970).

CHAPTER
XV

Polarization Response

15-1 INTRODUCTION

Although the scientific literature contains a great many references to polarized light, there are few books devoted totally to this subject [1, 2]. This can be partially attributed to the nomenclature problem that plagues radiometry in general. Polarized light is a tool that has found applications in many diverse fields resulting in numerous conventions.

Polarization is another degree of freedom in the characterization of target sources which can be employed to infer target attributes. Polarization is a fundamental property of electromagnetic radiation. The change that takes place in the polarization characteristics of electromagnetic radiation, as it interacts with materials and with electric and magnetic fields, is a convenient and accurate diagnostic tool.

All natural radiation is polarized to some extent (although unpolarized radiation can be generated). This characteristic of radiation causes a problem because of certain undesirable polarization properties of optical instruments. The response of optical spectrometers is considerably affected by polarization, although in suitably designed spectrometers polarization may be employed to increase the information yield [3].

The purpose of this chapter is to consider briefly the Stokes vector [3, p. 4] and various measurement techniques by which the polarization state of incoherent radiation can be completely characterized and to consider the effects of the undesirable polarization properties of electrooptical sensors on radiometric measurements.

15-2   POLARIZATION

There are various representations of the polarization of electromagnetic radiation. Recently, because of its dominant rǫle in its interaction with materials, the electric vector has become generally favored to describe polarization phenomena [4]. It was the polarization property of electromagnetic radiation that led historically to its characterization as a transverse wave. The representation of polarization that is the most useful is the one that unambiguously describes the behavior and permits the easiest solution of various problems encountered. Unfortunately, that may depend on the application [4, p. 5].

From the standpoint of classical physics, electromagnetic radiation consists of waves whose vibrations are transverse to the direction of propagation. *Polarized* radiation has a preference as to the vibrational pattern in a plane normal to the direction of propagation. The direction of vibration of the electric vector in the XY-plane can be specified as the *azimuth of vibration* ϕ as shown in Fig. 15-1.

The locus of the tip of the vector as it passes through the XY-plane is known as the *polarization figure*. When the vector remains fixed at any ϕ, the polarization is *linear*. When the vector E_ϕ rotates with time ($d\phi/dt > 0$), the polarization is *circular* or *elliptical*.

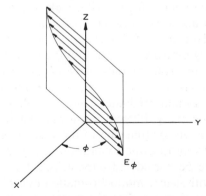

Figure 15-1   As the wave moves in time along the Z-axis, the electric vector oscillates in the XY-plane at the angle ϕ.

The linear type of polarization includes an infinite number of forms differing in azimuthal angle ϕ. The circular type of polarization has two forms which differ as to right or left handedness.

The elliptical type includes circular and linear as special cases. Elliptical polarization is illustrated in Fig. 15-2. For the general elliptical polarization the azimuth is given by the position ϕ of the major axis of the ellipse, the

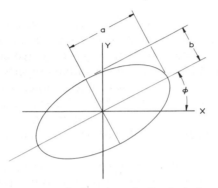

Figure 15-2   As the wave moves in time along the Z-axis, the locus of the tip of the electric vector in the XY-plane is an ellipse.

ellipticity is the ratio b/a where b is the semiminor axis and a the semimajor axis, and the handedness is *right* if the E-vector rotates in a clockwise direction and *left* if the E-vector rotates in the counterclockwise direction [4, p. 4]. Linear polarization corresponds to an ellipticity of zero, while circular polarization corresponds to an ellipticity of unity.

Depending on the field of study and experimental methods used, the sign of linear polarization and the handedness is defined differently.

Unpolarized radiation is difficult to illustrate graphically. The predominant characteristic of unpolarized radiation is that the vector has no preference as to the lateral direction of vibration or handedness. Radiation emitted by a single stationary dipole would be both monochromatic and polarized. Almost perfectly monochromatic radio waves are common, and they exhibit a high degree of polarization. However, optical radiation always has an appreciable bandwidth and originates from a large number of dipole radiators, and may include many different forms of polarization simultaneously. The resultant electric vector (the sum of the vectors originating from each dipole) may be as likely to exhibit vibration in one direction as any other and would therefore be considered unpolarized. Unpolarized radiation is therefore possible with *nearly* monochromatic optical radiation.

Most optical radiation, whether of natural or artificial origin, is neither completely polarized or completely unpolarized. Therefore the concept of the *degree of polarization* is important. The degree of polarization is defined as

$$P = (\Phi_p/\Phi_t) \times 100 \quad [\%], \tag{15-1}$$

where Φ_p represents the polarized part of the total flux Φ_t. However, a technique that can be readily implemented in the laboratory and which is based on a vibration-form dichotomy is given by

$$P = [(\Phi_{max} - \Phi_{min})/(\Phi_{max} + \Phi_{min})] \times 100 \quad [\%], \tag{15-2}$$

where Φ_{max} is the flux measured with a polarizer–analyzer adjusted to maximum transmittance and Φ_{min} the flux measured with the polarizer–analyzer adjusted to minimum transmittance. These forms are equivalent since the polarized component $\Phi_p = \Phi_{max} - \Phi_{min}$ and the total flux $\Phi_t = \Phi_{max} + \Phi_{min}$.

15-3 POLARIZERS AND RETARDATION PLATES

A *polarizer* is an optical device that, when supplied with unpolarized radiation, can produce a beam that is polarized. An *analyzer* is distinguished from a polarizer by the role it may perform: the production of polarized radiation (polarizer) or the detection of polarized radiation (analyzer).

There are at least four basic mechanisms for performing the function of the polarization of radiation: *dichroism, birefringence, reflection,* and *scattering.*

A *dichroic* polarizer preferentially transmits radiation of one polarization form and absorbs the orthogonally polarized form. A *linear* dichroic polarizer contains either long thin microcrystals or molecules that have been aligned with respect to their longest axis. The most common method to produce such a polarizer is to imbed such needle-shaped dichromophores in a plastic sheet and to stretch the sheet unidirectionally. The needles which had a random distribution tend to be turned approximately parallel with the stretch axis. The extent of absorption depends on the direction of the vibration of the electric vector with respect to the alignment of the dichromophores rather than the direction of propagation.

A *birefringent* material is an refractoanisotropic material, which means that the index of refraction of transmitted radiation depends on the direction of vibration of the electric vector. Such material exhibits a different index of refraction along two mutually perpendicular axes. Thus linearly polarized radiation will experience travel at a different speed and will be refracted in a different direction depending on the orientation of the radiation with respect to the axes. When nonpolarized radiation passes through such material two refracted rays are produced, each having its own direction of propagation and speed, and the two rays are found to be orthogonally polarized.

The radiation is propagated with the greatest speed along the axis exhibiting the minimum index of refraction (the speed being maximum in a vacuum), and is therefore referred to as the *fast axis*. Radiation propagated along the slow axis suffers a *relative* phase retardation. A birefringent material can therefore function as a *retarder*. When linearly polarized radiation passes through a retarder it is resolved into two components, one along the fast axis and one along the slow axis. The phase of one component is retarded relative to the other so that the two components recombine to form a beam of different polarization than that of the incident ray. The amount of retardation obtained depends on the wavelength of the radiation, the difference in the indices, and the pathlength. The most common retarders are quarter-wave (90°) and half-wave (180°).

A retarder has no effect on the polarization form of unpolarized radiation; however, a linear polarizer used in conjunction with a quarter-wave retarder will produce circularly polarized radiation. To do so the fast axis of the retarder must be oriented at 45° relative to the linear polarizer, and the radiation must pass through the linear polarizer first. In general, any form of polarization can be converted to any other form using appropriate linear polarizers and retardation plates. The effect of such devices upon a beam of polarized light can be determined by use of the Poincaré sphere or by the use of the Mueller or Jones calculus [1, p. 92].

Reflection polarizers preferentially reflect and transmit radiation of orthogonal polarization. Such reflection results from the interaction of the electric vector with the surface material. The relationships governing reflection from the surface of a dielectric material were worked out by Fresnel based on the boundary conditions. The Fresnel equations give the reflectance $\rho_\perp$ for radiation whose electric vector $E_\perp$ is *perpendicular* to the plane of incidence (the plane containing the beam and the normal to the surface) and $\rho_\parallel$ for radiation whose electric vector $E_\parallel$ is *parallel* with the plane as illustrated in Figs. 15-3 and 15-4. They are

$$\rho_\perp = \sin^2(\theta - \theta')/\sin^2(\theta + \theta') \tag{15-3}$$

and

$$\rho_\parallel = \tan^2(\theta - \theta')/\tan^2(\theta + \theta'), \tag{15-4}$$

where θ is the angle of incidence, θ' the angle of refraction, and θ'' is the angle of reflection.

For the special condition $\theta + \theta' = 90°$ the term $\tan 90° = \infty$ and the parallel reflectance $\rho_\parallel$ is zero. This special angle of incidence is called *Brewster's angle* or the *polarizing angle*, and is given by

$$\theta_B = \text{arc } \tan(n_2/n_1), \tag{15-5}$$

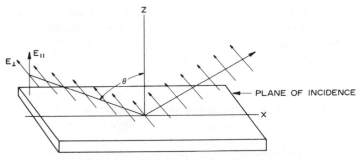

Figure 15-3   Polarization of a reflected ray.

where n_1 is the index of refraction for medium 1, and n_2 is the index of refraction for medium 2.

The degree of polarization for the reflected ray is given by

$$P = [(\rho_\perp - \rho_{||})/(\rho_\perp + \rho_{||})] \times 100 \quad [\%] \tag{15-6}$$

which for Brewster's angle is 100%. In practice the reflected ray is nearly 100% polarized and that is the basis of the application of polaroid sun glasses.

Numerous reflection polarizers are available that capitalize on these principles. Thin films deposited on a substrate [5, 6] and air-spaced "pile-of-plates" have been fabricated that function through the visible and IR region.

Scattered radiation of a Rayleigh atmosphere has been found to exhibit polarization characteristics and has been used to infer the vertical distribution of dust in the atmosphere [7].

When unpolarized radiation passes through a long slender slit, the emergent beam is partially polarized so that the electric vector tends to be lined up parallel with the slit [1, p. 85].

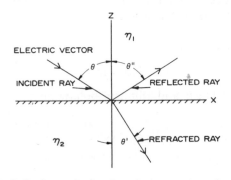

Figure 15-4   Reflection and refraction of a ray at a boundary surface.

Transmission gratings have been fabricated by embedding wires in a transparent substrate such as glass. Ordinary reflection gratings, such as those used in grating spectrometers, also exhibit polarization characteristics [8].

15-4   STOKES PARAMETERS

The state of polarization of a beam of radiant flux can be completely specified by giving the absolute value Φ_0 of the flux Φ, the degree of polarization P, the azimuth ϕ, the ellipticity b/a, and the handedness (see Fig. 15-2).

This can be accomplished in terms of the Stokes vector, which consists of a set of four entities called *Stokes parameters*. The vector, which consists of four physically real parameters, can be considered as a mathematical vector and used in conjunction with the Muller or Jones matrices [1, p. 109].

Incoherent radiation can be completely described by specifying the four parameters:

$$S_0^{\ 2} = S_1^{\ 2} + S_2^{\ 2} + S_3^{\ 2}, \tag{15-7}$$

where

$$S_0 = \Phi_0, \tag{15-8}$$

$$S_1 = S_0 P \cos 2\chi \cos 2\phi = S_0 P_1 \cos 2\phi, \tag{15-9}$$

$$S_2 = S_0 P \cos 2\chi \sin 2\phi = S_0 P_1 \sin 2\phi \tag{15-10}$$

$$S_3 = S_0 P \sin 2\chi = S_0 P_e. \tag{15-11}$$

Equations (15-8)–(15-11) are interpreted as follows: The first parameter S_0 gives the absolute value of the flux Φ_0; the second and third parameters, S_1 and S_2, give the degree of linear polarization P_1 and the azimuthal angle ϕ. The parameter S_3 gives the degree of elliptical polarization P_e and the ellipticity b/a.

The ellipticity is given by

$$b/a = \tan \chi, \tag{15-12}$$

and the degree of linear and elliptical polarization are each given, respectively, by

$$P_1 = P \cos 2\chi \tag{15-13}$$

and

$$P_e = P \sin 2\chi. \tag{15-14}$$

The inverse relationships are

$$\Phi_0 = S_0, \tag{15-15}$$

$$P = (S_1{}^2 + S_2{}^2 + S_3{}^2)^{1/2}/S_0, \tag{15-16}$$

$$\phi = \tfrac{1}{2}[\text{arc tan}(S_2/S_1)], \tag{15-17}$$

$$\chi = \tfrac{1}{2}[\text{arc tan}(S_3/(S_1{}^2 + S_2{}^2)^{1/2})]. \tag{15-18}$$

It is sometimes convenient to make use of the *normalized* Stokes parameters

$$s_i = S_i/S_0, \tag{15-19}$$

where $i = 1, 2, 3$ as a three-dimensional representation of the polarization state without considering the absolute flux Φ_0. To do so results in a simplification in determining the polarization parameters as is evident from the right-hand sides of Eqs. (15-9)–(15-11).

An instrument that measures all four Stokes parameters (or the three normalized parameters) is called a *complete polarimeter* or a *Stokes meter*.

15-5   MEASUREMENT OF STOKES PARAMETERS

There are a number of measurement techniques available to determine Stokes parameters. The compensator–analyzer–polarimeter (CAP), like other methods, requires that four measurements of flux be made with a polarization-insensitive detector. The CAP method uses various combinations of a compensator (usually a quarter-wave retarder) and an analyzer.

Continuous rotation of the elements of a CAP produces sinusoidal variations in the measured flux. The resultant modulated signal can be analyzed by Fourier techniques. The simplest rotating element polarimeter employs only a rotating analyzer (see Fig. 15-5). For this case the measured flux Φ_m is given by

$$\Phi_m = \tfrac{1}{2}[S_0 + S_1 \cos 2\phi + S_2 \sin 2\phi], \tag{15-20}$$

where ϕ is measured counterclockwise from the X-axis when viewing the source. The magnitude of the flux when viewed through a rotating analyzer is

$$\Phi_m = \Phi_0 \cos^2 \phi, \tag{15-21}$$

where Φ_0 is the absolute value of the linearly polarized flux. Actually, Eq. (15-20) can be solved using Eq. (15-15) for a simple two-measurement scheme where the flux is measured at $\phi = 0$ and 45°, to yield the values of S_1

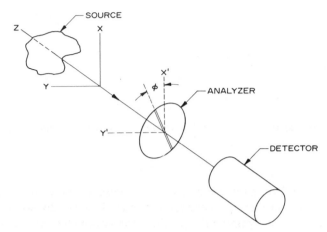

Figure 15-5 A simple rotating-element polarimeter.

and S_2. However, this is not a complete polarimeter since S_3 is not measured.

Actually, all four Stokes parameters can be found from the signal produced by rotating a single optical element. A complete polarimeter has been produced by rotating a compensator (retarder) with other fixed elements [9].

In another polarimetric method the components are physically stationary but the retardance is varied by either imposing changing electrical fields or stress forces upon a birefringent cell to modulate the retardance [10, 11].

Finally, interferometric polarimetry methods are attractive because retardation can be accomplished spatially in the two arms of the interferometer rather than in a birefringent retardation plate. A Michelson interferometer can be modified through the use of appropriate polarizers–analyzers to obtain high-resolution spectra of the Stokes parameters [12].

15-6 APPLICATIONS OF POLARIZATION

New applications of polarimetry are constantly being announced. It is not possible to do more than indicate some of the major areas.

Astronomical polarimetry provides information on a great variety of physical processes in astronomical objects. Of particular importance is the influence of magnetic fields on the polarization properties of light from stars, nebulae, and the sun.

A great deal of information can be obtained about atmospheric aerosols and Rayleigh scattering by observing the polarization state of light transmitted through planetary atmospheres. Zenith measurements of twilight airglow emissions, using photometers and radiometers, can be enhanced by preferentially discriminating against the scattered solar radiation which is highly polarized in a direction 90° to the sun.

Ellipsometry is a class of polarimetry that uses the polarization altering properties of an optical system. Optically active materials exhibit optical rotatory dispersion and circular dichroism due to the structure of the atoms in the molecule or crystal. An instrument designed specifically to measure the concentration of sugar solutions is known as a "saccharimeter." Some polarimeters can measure the rotation of the axis of polarization by an active substance to within 0.001°.

It is possible to detect and characterize a contaminant film on the surface of a solid even when the film is only a few angstroms thick using polarization techniques. This is surprising since the wavelength used for such measurements is about 6000 Å. The reason for this unique capability is that it capitalizes on the phase of reflection rather than the change in path difference as in interferometry.

The optical properties of materials are altered under the influence of external stresses such as mechanical stress or stress induced by electric or magnetic fields [13]. These properties can be characterized using polarization techniques.

Hundreds of other types of applications of polarizers and polarized radiation could be mentioned. The characteristics of polarized radiation provide a potential tool in target discrimination in remote sensing applications.

15-7   METROLOGICAL IMPLICATIONS OF POLARIZATION

As previously indicated, serious problems arise from undesirable polarization properties of certain optical sensors [14]. This is because the sensor response is considerably affected by polarization. To a great extent polarization effects have been ignored in many spectrometric measurements.

The assumption that objects are Lambertian and that polarization effects are not significant must be discarded in view of the evidence that radiation becomes polarized to some extent upon reflection. This is a problem in reflectance spectrometry because the measurement may be a function of sample orientation.

Absorption spectral measurements of optically active substances are subject to problems because the polarization is altered by the sample. The polarized component of the radiation is rotated in the case of linear polarization when passing through the sample. Different sensors will therefore measure different values of transmittance for such samples.

Any optical sensor that utilizes mirrors, slits, gratings, or beam splitters may exhibit some degree of polarization sensitivity. Many authors have reported that spectrometers are considerably affected by polarization [8, 14, 15]. Furthermore, the spectral responsivity of a spectrometer may be altered in a wavelength-dependent fashion.

Radiometers and photometers that use refractive lenses and antirefectance coatings, mirrors and high-reflectance coatings, or thin-film narrow bandpass filters are generally not polarization sensitive. This is because the optical components are usually circularly symetrical and the radiation falls at or near normal incidence upon these optical elements [16].

The concept of the degree of polarization can be used to characterize the polarization sensitivity of a sensor. The method of Eq. (15-2) or (15-20) can be applied, wherein a single linear polarizer is used to obtain both the degree of polarization P and the azimuth ϕ. This can also be accomplished by adjusting the polarizer to obtain a measure of the minimum and maximum flux, transmittance, or absorptance [15, p. 2229], in accordance with Eq. (15-2). Such measurements for typical optical spectrometers indicate that up to 30% wavelength-dependent polarization has to be expected in any spectrometer. This, of course, must be considered as a factor in determining the uncertainty in any absolute spectrometric measurements.

The subject of polarization of radiant sources has also been discussed in the literature but little is available on actual sources [15, p. 2231].

15-8   INFRARED POLARIZERS

Polarizers for the IR are available. However, most known materials that are birefringent in the visible tend to become opaque in the infrared. Furthermore, birefringent polarizers tend to be aperture limited [17].

Polarizers for the IR have been proposed that consist of variations of the reflectance pile-of-plates method. One of these uses four air-spaced germanium films inclined at the Brewster angle [18]; it works well over the range of 2 to 14 μm.

Brewster angle, pile-of-plates polarizers utilizing the transmitted beam meet the need for high quality, low cost, and large aperture, and exhibit small lateral beam displacement. Such polarizers have been fabricated out of commercial polyethylene plastic [6].

Such IR polarizers are adequate, in many applications, to determine the degree of polarization sensitivity of IR sensors and the degree of polarization of IR sources using the methods previously discussed.

REFERENCES

1   W. A. Shurcliff, "Polarized Light." Harvard Univ. Press, Cambridge, Massachusetts, 1962.
2   W. Swindell, ed., "Polarized Light." Dowden, Hutchinson & Ross, Stroudsburg, Pennsylvania, 1975.
3   P. S. Hauge, Survey of methods for the complete determination of a state of polarization. *Proc. Soc. Photo-Opt. Instrum. Eng.—Polarized Light* **88**, 3–10 (1976).
4   D. Clarke, Nomenclature of polarized light: Linear polarization. *Appl. Opt.* **13**, 3–5 (1974).
5   M. Ruiz-Urbieta, E. M. Sparrow, and P. D. Parikh, Two-film reflection polarizers: Theory and application. *Appl. Opt.* **14**, 486–492 (1975).
6   D. T. Rampton, and R. W. Grow, Economic infrared polarizer utilizing interference films of polyethylene kitchen wrap. *Appl. Opt.* **15**, 1034–1036 (1976).
7   Z. Sekera, and C. R. Nagaraja Rao, Skylight polarization: A criterion of atmospheric turbidity. *J. Opt. Soc. Am.* **55**, 595 (1965).
8   K. O. Lee and Y. P. Neo, Polarization irradiance distribution in the spectrum of infrared grating instruments. *J. Opt. Soc. Am.* **61**, 273 (1971).
9   F. Q. Orrall, A complete Stokes vector polarimeter. *In* "Solar Magnetic Fields" (R. Howard, ed.), pp. 30–36. Reidel, Dordrecht, Netherlands, 1971.
10 B. H. Billings, The electro-optic effect in uniaxial crystals of the dihydrogen phosphate type. *J. Opt. Soc. Am.* **42**, 12–20 (1952).
11 J. C. Kemp, Piezo-optical birefringence modulators: New use for long-known effect. *J. Opt. Soc. Am.* **59**, 950–954 (1969).
12 A. L. Fymat, Interferometric spectropolarimetry: Alternate experimental methods. *Appl. Opt.* **11**, 2255–2264 (1972).
13 K. Vedam, Applications of polarized light in materials research. *Proc. Soc. Photo-Opt. Instrum. Eng.—Polarized Light* **88**, 78–83 (1976).
14 K. Kudo, T. Arai, and T. Ogawa, Method for determining the degrees of polarization of infrared polarizers and monochromators. *J. Opt. Soc. Am.* **60**, 1046–1050 (1970).
15 F. Grum and L. F. Costa, Determination of polarization in optical instruments and its meteorological implications. *Appl. Opt.* **13**, 2228–2232 (1974).
16 W. W. Buchman, S. J. Holmes, and F. J. Woodberry, Single-wavelength thin-film polarizers. *J. Opt. Soc. Am.* **61**, 1604–1606 (1971).
17 L. Bergstein, Novel thin-film polarizer for the visible and infrared. *J. Opt. Soc. Am.* **61**, 665 (1971).
18 J. M. Bennett, D. L. Decker, and E. J. Ashley, Infrared germanium-film polarizer. *J. Opt. Soc. Am.* **60**, 1577 (1970).

CHAPTER

XVI

Practical Calibration of Cryogenic LWIR Systems

16-1 INTRODUCTION

The practical calibration of cryogenic LWIR systems includes all the basic problems of the calibration of optical sensors in general and those of vacuum and cryogenic operation in particular. Therefore, the application of the theory to the problem of calibration given here is relatively general except perhaps for the obvious complications associated with vacuum-cryogenic technology. The calibration consists of two phases:

(1) an engineering or optical bench test calibration, and
(2) a final calibration in a suitable cold chamber.

The engineering calibration might be the final step in the fabrication of a system, or of a performance evaluation test of an existing system, and provides documentary proof of performance. The major compromises involved in the engineering calibration of a cryogenic sensor, as visualized here, are that this calibration requires several extrapolations to derive either an extended-area source calibration or a distant small-area calibration, and that insufficient data are obtained to provide an estimate of the uncertainty. Consequently, the uncertainty in the resultant responsivities is rather high,

163

perhaps a factor of 2, which may be unsatisfactory for measurements but quite satisfactory to establish instrument performance.

There are two factors that distinguish the final calibration. First, the calibration is designed to approximate, as far as possible, the conditions under which the measurements are to be made so that the source can be qualified independently of the instrument parameters. This includes factors such as full collector illumination, similar background conditions, and similar target geometry and power spectral density characteristics. Second, sufficient data must be collected to provide for an estimate of the calibration uncertainty. The final calibration will be accomplished in accordance with these factors as far as it is economically and practically possible.

16-2   ENGINEERING CALIBRATION

An illustration of a high-vacuum cryogenic IR radiometer fitted with a cold pinhole cover for an engineering calibration is shown in Fig. 16-1. The entire optical subsection: preamplifier, detector, lenses, chopper, filter, and baffle are cryogenically cooled by thermal conduction from the Dewar cold finger. This eliminates any background flux that might originate within the sensor and results in enhanced detector performance [1].

These IR systems are frequently used to make *in situ* or exoatmospheric measurements of emissions of the upper atmospheric airglow and aurora [2]. The sensor, lifted aloft on a rocket carrier, is exposed to the atmospheric emissions only after it has been carried to an altitude such that the ambient vacuum environment will not compromise the cryogenic performance. The cover is then jettisoned, exposing the collector full field to the optically thin atmospheric radiant emissions against the cold background of outer space. Consequently, such sensors normally exhibit excessive response to 295K environments.

The normal procedures of an optical bench test calibration can be performed only under the conditions that the vacuum be maintained and that the ambient radiation levels be reduced below saturation levels. This is accomplished, as shown in Fig. 16-1, by using an ambient temperature IR window to maintain internal vacuum and a cold pinhole aperture to reduce the sensor response to the ambient 295K background to levels sufficiently below saturation to permit a limited calibration [3]. Sometimes it is necessary to observe the preamplifier output signals when the sensor signal conditioning amplifiers remain saturated.

Considerable work should be done before the engineering calibration begins and the following items must be designed and fabricated:

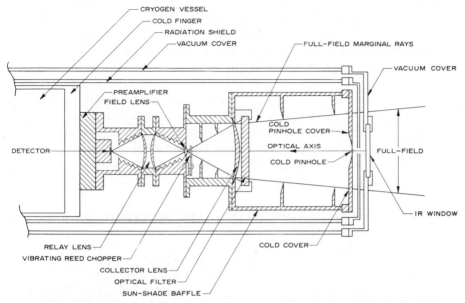

Figure 16-1 Illustration of a high-vacuum cryogenic IR system fitted with a cold pinhole cover for an engineering calibration.

(1) A cold pinhole cover must be designed, fabricated, and fitted to the instrument aperture as illustrated in Figure 16-1.

(2) A cold cover, similar to the pinhole cover, must be fabricated to prevent any radiant energy from entering the aperture. This cover is used to provide zero background readings.

(3) A vacuum cover containing an IR window must be designed, fabricated, and mounted as shown in Fig. 16-1. The IR window must be made of suitable optical materials to provide optical transmission over the spectral range of the sensor.

The instrument specifications are used to calculate the expected response to the ambient environment under the conditions of full collector illumination and of pinhole illumination. The size of the pinhole is designed such that the ratio of the pinhole area to effective collector area will achieve the desired reduction in the sensor response to the ambient background.

The first test is performed with item (2), the cold, dark, aperture cover, in place. The sensor is vacuum pumped and cryogenic cooling commenced. The detector response and internal temperature monitors are recorded as a function of time. The normal cool-down performance pattern is thus documented permitting early detection of a malfunction due to inadequate vacuum, thermal short circuits, etc.

Usually the detector will cool and become operational some time before the optics, so that the cooling rate of the optics can be observed indirectly by the detector response to the thermal radiations emanating from the mirrors, baffles, etc. This response to the thermal radiations of the optical components, which become coded by the optical filter, is by definition the background referred to in this test. This background yields an output signal. As the instrument cools, the background signal should go to zero so that the only output is random noise with zero mean. Failure of the background signal to reach zero is caused by either a light leak in the optical system or a component part that is inadequately heat sinked. This phase of engineering calibration may reveal a need for design changes in the heat-sinking of component parts or light seals and often uncovers problems that are difficult to solve.

The noise output is analyzed as outlined in Chapter VIII so that the presence of any dc offset or coherent noise forms can be identified. The results of the dark-noise analysis should be filed in the calibration report.

The second test consists of attaching the pinhole and IR window to the instrument with the alignment on the optical axis. The instrument is again vacuum pumped and cooled. This time the output will not go to zero, but will level off at a value corresponding roughly to a 295K blackbody source. This response to ambient background must be subtracted along with dc offset errors from the response to the blackbody standards.

The sensor may be mounted on the optical bench as illustrated in Chapter XI. The calibration of the pinhole-modified sensor can proceed as appropriate. The results of this calibration must be extrapolated to the full-collector case giving careful considerations to the effects of the pinhole on all the instrument parameters.

16-3   FINAL CALIBRATION

There are both practical and theoretical limits beyond which it is impossible to simulate actual measurement conditions using laboratory standards. To begin with if the field conditions were perfectly known, it would not be necessary to make any measurements. In addition the target may not fit the idealization of being either a point source or an extended source. Finally, the spectral distribution of energy in typical blackbody standards may be vastly different than the field target and often poses more severe problems of spectral purity. These problems are greatly compounded in the calibration of high-vacuum cryogenic LWIR sensors.

The major considerations having practical value are as follows:

The target background.   The target may be a star measured against the scattering and emitting characteristics of the atmosphere or it may be the emissions of the atmosphere measured against the galactic background. What is one person's target may be another person's background. However, there is a significant difference between the ambient background that radiates like a 295K blackbody and the background of outer space. Chambers that simulate outer space, that provide cryogenic cooling and hard vacuum conditions, are expensive to fabricate, operate, and maintain. Such chambers vary in size from small bolt-on adapters to large complex facilities [4].

The target geometry.   The target geometry is often well known. Either an extended area source or a distant small area source must be used as a best approximation of the target characteristics.

An extended-area source calibration of a cryogenically cooled LWIR spectrometer can be accomplished by coupling the instrument to a cryogenically cooled chamber that contains a large area, temperature-controlled surface [4]. This surface, properly serrated and blackened, serves as an extended area blackbody source. The chamber must provide cryogenic baffling and vacuum continuity with the sensor. Such a cryogenically cooled source provides an effective zero-level background. The source temperature can then be increased in steps to provide an output throughout the entire dynamic range of the instrument yielding data that can be analyzed for absolute calibration and spectral purity. This method, using as it does an extended area source, suffers from the problems associated with the long wavelength shift of the radiation as described in Chapter XIII, and is therefore not very useful at wavelengths below about 10 μm. The extended area source often saturates the sensor when the temperature is increased enough to overcome this long wavelength shift of radiation.

An alternative method has been proposed [5] that makes use of a cryogenically cooled integrating sphere as an extended area greybody source. The energy admitted into the integrating sphere from a relatively hot blackbody can be suitably attenuated by limiting the source area.

A distant, small area, source calibration can similarly be obtained using cryogenically cooled chambers that contain sources, collimators, and filter and aperture sets. Several large military and commercial chambers are available such as the Aerospace Research Chamber (7V) and the Mark VI Chamber of the Von Karman Gas Dynamics Facility, Arnold Air Force Base, Tennessee [6, 7]; the Advanced Sensor Evaluation and Test Facility ASET of the McDonnell Douglas Astronautics Co., Huntington Beach, California [8]; and the Low Background Calibration Facility LBCF of the Honeywell Inc., Radiation Center, Lexington, Massachusetts.

The difficulties associated with the measurement of the far-field spatial response of the field of view, as considered in Chapters X and XI, are compounded. The cold chambers, like any facility, are limited by scattering off the walls. In some cases it may be appropriate to experimentally evaluate the field of view of cryogenic IR sensors under ambient conditions by substituting an ambient detector of similar geometry, provided that theoretical calculations can give an extrapolation from ambient to cold conditions. Such measurements have been made at 10 μm using a CO_2 laser to extend the dynamic range above the ambient level.

The target-relative spectral power density.   The only practical condition under which a measurement can be made independently of the relative spectral responsivity of the instrument is when the calibration source and the target have the same spectral power density curve. This condition can be achieved when the target is known to radiate as a blackbody at a known temperature. This is a rather limited case; generally it is difficult to simulate the source spectral properties even if they are known, with a radiation standard. The major problems are the long wavelength shift of the radiation of low-temperature blackbody sources and the difficulty of operating and qualifying standard sources, monochromators, detectors, etc., under conditions of hard-vacuum cryogenic temperature operation.

REFERENCES

1   W. L. Wolfe, ed., "Handbook of Military Infrared Technology," pp. 515–516. Off. Nav. Res., Washington, D.C., 1965.
2   A. T. Stair, Jr., J. C. Ukwick, D. J. Baker, C. L. Wyatt, and K. D. Baker, Altitude profiles of infrared radiance of O_3 (9.6 μm) and CO_2 (15 μm). *Geophys. Res. Lett.* **1**, 117–118 (1974).
3   T. P. Condron, "Calibration of a Liquid Helium Cooled CVF Radiometer in a Warm Environment," Sci. Rep., AFCRL-TR-73-0480, Project 8692, Opt. Phys. Lab., Air Force Cambridge Res. Lab., L. G. Hanscom Field, Bedford, Massachusetts (1973).
4   C. L. Wyatt, Infrared spectrometers: Liquid-helium-cooled rocketborne circular-variable filter. *Appl. Opt.* **14**, 3086–3091 (1975).
5   T. P. Condron, R. P. Heinesch, and D. J. Lovell, Integrating sphere performance in the 0.3 to 8 μm spectral region. *Proc. Soc. Photo-Opt. Instrum. Eng.—Mod. Util. Infrared Technol.* **62**, 135–150 (1975).
6   F. G. Sherrell, A general purpose infrared source for sensor tests at 20°K background temperature. Iris Paper No. B-4, *Stand. Spec. Group, Nav. Electron. Lab. Cent., San Diego, Calif., 1974.*
7   F. Arnold, and F. W. Nelms, "AEDC Long Wavelength Infrared Test Facilities," AEDC Tech. Rep. AEDC, Arnold Air Force Stn., Tennessee (1975).
8   R. H. Meier and A. B. Dauger, Spectral calibration of infrared space sensors. *Proc. Soc. Photo-Opt. Instrum. Eng.—Long Wavelength Infrared* **67**, 103–110 (1975).

XVII

Calibration of a Radiometer—
A Detailed Example

17-1 INTRODUCTION

This chapter presents a review of the basic philosophy of calibration as applied to the specific case of a radiometer. The techniques and the associated calibration equipment are described and the computer summaries are given. It is intended that this chapter provide a demonstration of the application of the calibration theory to the real problem of calibration in such a way as to provide a "here is how you do it" step-by-step procedure.

It should be realized that the details will vary in such a calibration exercise depending on the type of sensor and the type of calibration equipment available. Actually, calibration technology will probably always be developmental; that is, the procedures used will evolve as new equipment and technology become available. Unfortunately, the techniques used in any situation depend not only on the state-of-the-art but also on the personnel and monetary resources available.

17-2 OPERATIONAL PROCEDURES

The final product of the calibration of a radiometer is a mathematical equation (the calibration equation) which gives the desired flux Φ in terms of the instrument output Γ. The form of the calibration equation is

$$\Phi = a\Gamma + b\Gamma^2 \tag{17-1}$$

which is of second degree and for which the intercept is zero. This equation is used for systems that are nominally linear but that typically exhibit some degree of nonlinearity for large signal levels. There are three reasons for forcing the function to pass through the point $\Gamma = 0$ for $\Phi = 0$. The first is that a nonzero intercept results from an "offset error" which may change with environment and history. The second is that when the data from a multichannel sensor are projected onto the high-gain channel, as in the linearity analysis, the offset error is multiplied by the relative gain. The third, and most important, is that an offset error obscures spectral impurities which have a similar appearance. For these reasons, the offset is measured and subtracted from all data before processing.

The Utah State University radiometer was designed to make extended source measurements of high-altitude emissions of the atmosphere from a rocketsonde carrier. An extended area source calibration satisfies the rule of good performance in this case. However, the distant small area source calibration provides for an evaluation of the system linearity and field of view and for a cross check on the extended area source absolute calibration. Thus the recommended procedure is to calibrate sensors on both facilities whenever possible and then compare the results. The radiometer is shown in Fig. 17-1 and its specifications are given in Table 17-1, where it is noted that two output channels, with a gain difference of approximately 100, are available. This provides for an extended dynamic range using linear output channels.

Figure 17-1 Illustration of the Utah State University radiometer.

TABLE 17-1

Specifications for the Utah State University Radiometer[a]

Electronics

Signal conditioning amplifiers: ac
Bandwidth: 10 Hz
Signal channels: 2 linear, gain difference 100
                             1-chopper reference (sine wave)
  Levels: ± 10.5 V at $Z_0 \leq 1$ kΩ
Temperature monitors: preamplifier, detector, and optics
  Levels: ± 10.5 V at $Z_0 \leq 1$ kΩ
DC restoration: phaselock
Bandwidth: 1 Hz

Optical

Optics: 2 lens coated silicon miniscus
Aperture: 0.59-in. (1.50-cm) diameter
Field of view: 4×4 deg^2 (4.9×10^{-3} sr)
Baffle: sunshade
  Rejection: 10^{-12} at 35° off-axis
Filter: interference
  Wavelength: 4.3 μm
  Half-power bandwidth: 0.25 μm

Radiometric

Detector type: InSb
Chopping frequency: 54 Hz
Noise equivalent sterance [radiance]: 1×10^{-11} (W cm^{-2} sr^{-1}) at 4.3 μm (design goal)
Cryogen: LN$_2$

---

[a] The units used for length, in the specifications and calibration data, are centimeters rather than meters in order to satisfy the customer (user) requirements.

Two data entry forms have been developed to organize and systematize the data collection. The first for the distant small area source is shown in Fig. 17-2, and the second for the extended area calibration source is shown in Fig. 17-3. The data were recorded on these forms as the calibration of the radiometer proceeded.

17-2-1   The Distant Small Area Source Calibration

The distant small area source consists of a collimator system (mirrors, filter wheel, aperture wheel, and baffles) which can be cooled with liquid nitrogen or liquid helium when necessary. The cryogenic radiometer must be

Form 1.   DISTANT SMALL-AREA SOURCE CALIBRATION CC-75-1
                        RADIOMETRIC CALIBRATION

Sensor Model No. _usu Rad_ Wavelength _4.3μm_ Calibration Date _7 Nov 1976_

1. Dark Noise Data

Channel Designation	Offset Voltage	Relative Gain
Hi	0.010	1.0
Lo	0.001	102.88

2. Field of view survey   x = _0.225°_   y = _0.200°_

3. Linearity Data

Aperture No.	Output Voltage					
	Ch. _Lo_	Ch. _hi_	Ch._____	Ch._____	Ch._____	Ch._____
1	0.025	2.89				
2	0.058	6.23				
3	0.123	sat.				
4	0.234					
5	0.448					
6	1.000					
7	1.602					
8	3.201					
9	6.460					
10	8.220					

4. Absolute Calibration Data

Test No.	Blackbody			Output Voltage					
	Dial	°C	°K	Ch. _Lo_	Ch._____	Ch._____	Ch._____	Ch._____	Ch._____
1	270	230	503	2.49					
2	310	260	533	3.48					
3	340	285	558	4.60					
4	380	315	588	6.20					
5	405	340	613	7.52					
6	425	355	628	8.50					
7	440	370	643	9.62					
8									
9									
10									
11									
12									
13									
14									
15									
16									
17									
18									

5. Field of View Data -- Near Field   ✓

6. Field of View Data -- Far Field   ✓

7. The spectral bandpass calibration   ✓

Figure 17-2 Form 1—to aid in data collection for distant small-area calibration.

Form 2.   EXTENDED-AREA SOURCE CALIBRATION CT-69
                      RADIOMETRIC CALIBRATION

Sensor Model No. _usu Rad._ Wavelength _4.3 µm_ Calibration Date _4 Nov. 1976_

Test	Blackbody No. 2			Output Voltage					
No.	Dial	°C	°K	Ch. _Lo_	Ch. _hi_	Ch.	Ch.	Ch.	Ch.
1	538.8	-114.5	158.5	0.009	1.000				
2	561.6	-108.5	164.5	0.020	2.000				
3	585.8	-102.7	170.3	0.038	4.000				
4	692.2	-96.3	176.7	0.079	8.22				
5	628.6	-92.3	180.7	0.120	Sat.				
6	674.1	-81.2	191.8	0.352					
7	724.5	-68.8	204.2	0.99					
8	778.1	-55.6	217.4	2.664					
9	805.0	-49.0	224.0	4.050					
10	852.0	-37.3	235.7	8.070					
11									
12									
13									
14									
15									
16									
17									
18									
19									

Figure 17-3 Form 2—to aid in data collection for extended-area source calibration.

interfaced with the collimator to provide vacuum, cold shield, and optical baffling continuity.

The radiometer and the collimator are mated, vacuum pumped, and cooled prior to calibration. This operation usually requires about 48 hr. Thereafter, the calibration proceeds as follows:

1. The dark-noise test. The aperture wheel was placed at a closed position to block any radiant flux that might have reached the radiometer aperture. The offset voltage was measured with an integrating voltmeter and entered on Form 1 (Fig. 17-2). An oscillogram of the high-gain channel output noise was recorded, as shown in Fig. 17-4, from which the system noise standard deviation was estimated at $\sigma = 16$ mV. The gain difference between high-gain and low-gain channels was previously measured electronically and recorded on Form 1 at 103.

Figure 17-4   An oscillogram of the output noise of the Utah State University radiometer.

2.   The field of view survey.   This was accomplished by setting the aperture wheel and blackbody to yield a convenient output signal. The collimator mirror was set at $Y = 0$ and swept through X. The midpoint was noted at $X = 0.225°$. The procedure was repeated with the mirror set at the X-axis midpoint and swept through Y, and the Y-axis midpoint value was noted at $Y = 0.200°$. The general features of the field-of-view response were found to agree with the sensor specifications. The collimator was set at the field-of-view midpoint for the linearity and absolute calibration tests which follow.

3.   The linearity test.   Vignetting cannot occur because the field of view of this radiometer is 4° (full angle) while the maximum divergence in the collimator beam is only 0.3°. It is preferred to evaluate the linearity by observing the output as a function of source area, at constant blackbody temperature, to eliminate the possibility of spectral impurity problems. The largest aperture was used for initial adjustments and the blackbody temperature was set to yield an output signal on the low-gain channel that was just under saturation. The output voltage was measured with an integrating voltmeter, to reduce the effects of system noise, and recorded on Form 1 for each of the apertures in the set.

4.   The absolute calibration test.   It is preferred to evaluate the absolute responsivity by observing the output as a function of source temperature, at constant aperture area, to test the absolute data for spectral purity. Long-wavelength leakage will result in a responsivity that is an inverse function of source temperature, and it can be identified, when present, by a rather large standard deviation of the quality-of-fit of the data points to the transfer function equation.

Long-wavelength leakage is minimized, in the absolute calibration, by using the smallest aperture, the lowest gain channel, and the highest blackbody temperatures possible: this provides maximum energy at shorter wavelengths.

The aperture was set at a convenient position to produce a relatively small signal on the low-gain channel (where the noise is minimum). The temperature was then increased in relatively small steps to yield a number of points until the output approached full scale. It is important to allow the blackbody to stabilize between each test.

5. *Field of view—near field.* The near field was measured as follows: The aperture was set to a small size to provide a divergence in the collimated beam that is very small compared with the full field. The blackbody temperature was adjusted so that the peak (on-axis) response was near full scale on the low-gain channel. The mirror was set at an arbitrary value of Y and scanned through X. Then Y was incremented in steps equal to about 1/10th of the full field. The data were recorded on a strip chart recorder and read off later using a reader card punch machine.

6. *Field of view—far field.* The far field was measured as follows: The aperture and blackbody temperature were adjusted so that the output was near full scale on the low-gain channel. The mirror was scanned through X with Y set at the midpoint. The output for this single scan was recorded for both high-gain and low-gain channels to obtain the maximum dynamic range.

7. *The spectral bandpass calibration.* The instantaneous spectral bandpass function $R(\lambda)$ of the radiometer was measured to provide a calculation of the normalized band sterance [radiance]. This required the use of a monochromatic source of flux and a standard detector. The filter transmittance cannot be used because it is modified by the spectral detector responsivity, internal (system) reflections, and nonperpendicular incident flux. The monochromator was used in conjunction with the collimator in the same manner as the blackbody source was used in the previous calibrations.

17-2-2 The Extended-Area Source Calibration

The extended-area calibration source contains a temperature-controlled plate that has been serrated and painted black to serve as an extended area source. The radiometer and extended area source must be interfaced so as to provide vacuum, cold shield, and optical baffling continuity. After they were mated, vacuum pumped, and cooled, the extended area calibration commenced as follows: The temperature was varied in steps to produce radiometer outputs from zero to full scale. The output voltage was measured with an integrating voltmeter (to reduce the effects of system noise) and recorded on Form 2 (Fig. 17-3) for each temperature.

17-3   CALIBRATION SUMMARY

This radiometer was calibrated on both the distant small area source and the extended area source. The details are given in Form 1 and 2 (Figs. 17-2 and 17-3), computer print-outs (Tables 17-2–17-7), and graphical plots (Figs. 17-5 and 17-6). The data were processed using the algorithms given in the appropriate chapters of this book.

TABLE 17-2

Linearity Data Set, Distant Small-Area Source

AREA	VOLT	CHANNEL
2.927−04	2.890	HI (1)
6.300−04	6.230	HI (1)
2.927−04	.025	LO (1)
6.300−04	.058	LO (1)
1.241−03	.123	LO (1)
2.435−03	.234	LO (1)
5.082−03	.448	LO (1)
9.934−03	1.000	LO (1)
2.034−02	1.602	LO (1)
3.986−02	3.201	LO (1)
8.173−02	6.460	LO (1)
9.931−02	9.220	LO (1)

TABLE 17-3

Quality of the Fit of the Linearity Data to the Standard Nonlinear Response Equation and the Measure of the System Nonlinearity, Distant Small-Area Source

STANDARD DEVIATION (PERCENT)	FULL SCALE NONLINEARITY (PERCENT)
7.565	21.076

```
THE RELATIVE TRANSFER FUNCTION IS
    AREA =  1.0505−04V  +   2.2140−08V**2
PROJECTED ONTO THE HI (1) GAIN CHANNEL
    FULL SCALE =  1000.00V
```

TABLE 17-4

Quality of the Fit of the Absolute Radiant Flux Data to the Standard Nonlinear Response Equation, Distant Small-Area Source

WAVELENGTH LAMBDA (MICRO M)	WAVENUMBER (1/CM)	STANDARD DEVIATION (PERCENT)
4.300	2325.581	2.129

TABLE 17-5

Band Sterance [Radiance] Calibration Equations $(W/(SQCM-SR))$, Distant Small-Area Source

CHANNEL	EQUATION
1 HI	$L = 1.240-09V + 2.613-13V**2$
1 LO	$L = 1.276-07V + 2.766-09V**2$

TABLE 17-6

Quality of the Fit of the Absolute Radiant Flux Data to the Standard Nonlinear Response Equation, Extended-Area Source

WAVELENGTH LAMBDA (MICRO M)	WAVENUMBER (1/CM)	STANDARD DEVIATION (PERCENT)
4.300	2325.581	2.853

TABLE 17-7

Band Sterance [Radiance] Calibration Equations $(W/(SQCM-SR))$, Extended-Area Source

CHANNEL	EQUATION
1 HI	$L = 1.581-09V + 2.429-13V**2$
1 LO	$L = 1.627-07V + 2.571-09V**2$

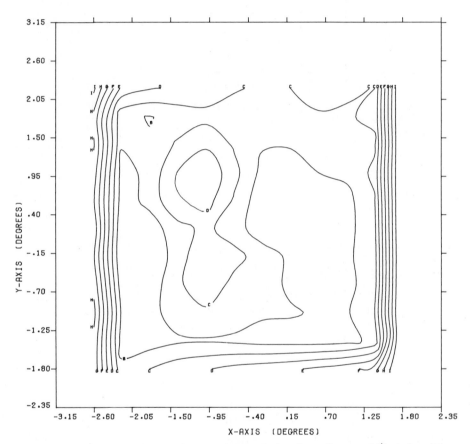

Figure 17-5 Near-field plot of the Utah State University radiometer field of view. (The contour intervals are $A = 1.0$, $B = 0.9$, etc.)

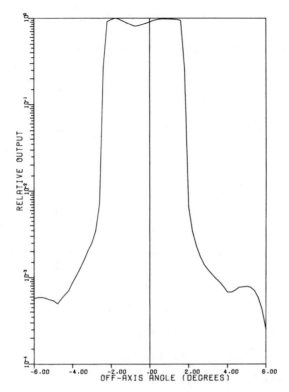

Figure 17-6   Far-field plot of the Utah State University radiometer field of view.

17-3-1   Sterance [Radiance]

The absolute sterance [radiance] of an extended-area source is obtained using the calibration equations as

$$L = 1.58 \times 10^{-9}V + 2.43 \times 10^{-13}V^2 \quad [\text{W cm}^{-2}\text{ sr}^{-1}] \qquad (17\text{-}2)$$

for the high channel and

$$L = 1.63 \times 10^{-7}V + 2.57 \times 10^{-9}V^2 \quad [\text{W cm}^{-2}\text{ sr}^{-1}] \qquad (17\text{-}3)$$

for the low channel (see Table 17-7), where V is the output voltage.

17-3-2   Dynamic Range

The standard deviation (rms value) of the high channel dark noise is estimated at 16 mV (see Fig. 17-4). Substitution of this value into Eq. (17-2) yields the noise equivalent sterance [radiance] NES [NER]

$$NES = 2.53 \times 10^{-11} \quad [W \ cm^{-2} \ sr^{-1}] \qquad (17\text{-}4)$$

for high channel. Full-scale output corresponds to approximately 10 V on low channel

$$L_{max} = 1.89 \times 10^{-6} \quad [W \ cm^{-2} \ sr^{-1}]. \qquad (17\text{-}5)$$

17-3-3   Calibration Uncertainty

The calibration equations are derived from the raw data by best-fit techniques. A measure of the calibration uncertainty is obtained as the standard deviation of the quality of fit as

$$\sigma_L = \pm 7.6\% \quad \text{(linearity: Table 17-3)}, \qquad (17\text{-}6)$$

$$\sigma_W = \pm 2.9\% \quad \text{(absolute: Table 17-6)}, \qquad (17\text{-}7)$$

$$\sigma_N = 0.016 \quad \text{V (system noise: Figure 17-4)}, \qquad (17\text{-}8)$$

and the uncertainty of the standard source is

$$\sigma_S = 14.2\% \qquad (\pm 1K \ at \ 158K). \qquad (17\text{-}9)$$

The precision (repeatability) of the sensor is

$$\sigma_P = (\sigma_N^2 + \sigma_W^2 + \sigma_L^2)^{1/2}, \qquad (17\text{-}10)$$

which is 8.1% for large signal-to-noise ratios (where σ_N can be neglected). For signals near the threshold of sensitivity, the precision is dominated by system noise.

The accuracy of the sensor is estimated by combining the precision and standard source uncertainty as

$$\sigma_A = (\sigma_P^2 + \sigma_S^2)^{1/2}, \qquad (17\text{-}11)$$

which is $\pm 16.4\%$ for large signals.

This analysis is based primarily on the extended area source absolute calibration. It is noted that an absolute calibration was also obtained from the distant small area source data. The calibration equations for this calibration (see Table 17-5) differed from those of the extended area source calibration by approximately 22%.

XVIII

Calibration of an Interferometer–Spectrometer— A Detailed Example

18-1 INTRODUCTION

Chapter XVII presented a detailed step-by-step procedure for the calibration of a radiometer. The absolute calibration of a spectrometer is accomplished by generally following the same basic procedure. This chapter emphasizes only those aspects of spectrometer calibration that differ significantly from those of radiometric calibration; therefore, this chapter should be read with a clear understanding of the material in Chapter XVII.

The absolute calibration of a spectrometer is roughly equivalent to the calibration of n radiometers where n is the number of spectral resolution elements in the free spectral range. Consequently, the quantity of data that must be processed is voluminous and this dictates on-line computer processing.

Most spectrometers are polarization sensitive and most sources are polarized to some degree. This is especially true of grading spectrometers and of interferometer spectrometers (see Chapter XV). Mirrors, slits, and gradings tend to polarize the light reflecting from or passing through such devices. Therefore, the absolute responsivity obtained in a calibration exercise is a function of the physical orientation of the spectrometer with respect to the source. Basically the transmittance of the various optical components in the spectrometer depends on the polarization of the source. Reports in the literature [1, 2] indicate that optical spectrometers typically exhibit a wavelength-dependent polarization sensitivity of up to 30%. Apparently the effects of polarization on spectrometery have largely been overlooked and there exists a need for information on the polarization characteristics of sources and spectrometers. Finally, the additional complexity and requirements for greater precision and stability of the components of a high-resolution spectrometer suggest that these instruments are more subject to errors.

Recognition of these problems leads to the concept that spectrometers and radiometers should be implemented in pairs—spectrometers to provide the *shape* of the flux and radiometers to measure the *absolute* band flux. In spite of these difficulties it is often desirable (or even necessary) to provide an absolute calibration of a spectrometer although the uncertainty may be difficult to estimate.

18-2   OPERATIONAL PROCEDURE

The Utah State University field-widened interferometer (FWIR), which is shown in Fig. 18-1, was designed to provide measurements of the extended area, night-sky emissions from a ground observatory [3, 4]. The specifications of this interferometer vary depending on the type of detector employed.

Only the distant small area source calibration is reported here to emphasize those aspects of spectrometer calibration that differ from those of radiometer calibration.

The interferograms were recorded on analog magnetic tape and later digitized and transformed by fast Fourier transformation (FFT). The computer output was formatted into a catalog of all the calibration spectrograms appropriately identified. Figure 8-1 gives one page of the spectrogram catalog and illustrates a number of scans that were used to obtain a linearity analysis. The scans, as recorded in the catalog, were visually selected for analysis; then the computer was used to calculate the dark-noise standard deviation and to tabulate the data appropriate for calibration analysis.

The linearity analysis and the absolute calibration proceed essentially as with a radiometer, except that for each setting the output consists of a spectrum rather than a single data point. Different from a radiometer, it is

Figure 18-1 Illustration of the Utah State University field-widened interferometer.

also necessary to obtain a spectral scan position calibration, which provides the wave number as a function of percentage of scan. This is accomplished through the use of a neon–argon lamp source which provides known emission bands throughout the free spectral range of the interferometer. Figure 18-2 shows the spectrum of the neon–argon lamp where the wave numbers of several significant lines have been identified.

The linearity data was obtained, as before, by varying the apertures while holding the temperature constant to avoid spectral purity problems. The data was chosen at the particular wave number that exhibited the greatest

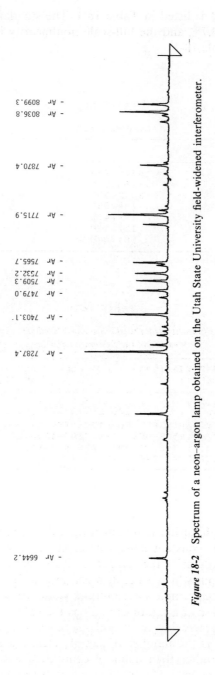

Figure 18-2 Spectrum of a neon–argon lamp obtained on the Utah State University field-widened interferometer.

dynamic range and is listed in Table 18-1. The standard deviation of the quality of fit was 4.7% and the full-scale nonlinearity is less than 1% for the FWIR (see Table 18-2).

TABLE 18-1

Linearity Data Set

AREA	VOLT	CHANNEL
3.986-02	6394.720	PREAMP
2.034-02	3050.610	PREAMP
9.994-03	1472.590	PREAMP
5.082-03	744.950	PREAMP
2.435-03	392.430	PREAMP
1.241-03	197.370	PREAMP
6.300-04	101.040	PREAMP

TABLE 18-2

Quality of the Fit of the Linearity Data to the Standard Nonlinear Response Equation and the Measure of the System Nonlinearity

STANDARD DEVIATION (PERCENT)	FULL SCALE NONLINEARITY (PERCENT)
4.707	-.692

THE RELATIVE TRANSFER FUNCTION IS
    AREA = 6.4536-06V + -6.9103-12V**2
PROJECTED ONTO THE PREAMP GAIN CHANNEL
    FULL SCALE = 6463.54V

The absolute calibration is obtained (as with a radiometer) by varying the temperature, at constant source area, in order to provide a test for spectral purity. Ordinarily the data is digitized at a relatively high rate resulting in an output which is nearly a continuous function of wavelength. Then the data at each sample wave number would be subjected to a best-fit analysis as in the case of a radiometer. Thus the calibration of a spectrometer requires the processing of roughly n times as much data as a radiometer (where n is the number of sample elements in the free spectral range). However, rather than define n calibration equations, the relative inverse spectral responsivity function is calculated and used in conjunction with a single-calibration equation as given in the next section. The standard deviation of the quality of fit and the relative inverse spectral responsivity function are normally plotted to provide a continuous representation of those functions through the free spectral range. For purposes of illustration, only selected wavelengths are processed in this example. The relative inverse spectral responsivity function is generated in accordance with the algorithms given in Chapter XIII.

18-3   CALIBRATION SUMMARY

The FWIR was calibrated on the distant small area source. Only those details that vary significantly from that of a radiometer are reported here. The details are given in Figs. 8-1, 18-1, and 18-2 and Tables 18-1 through 18-5.

TABLE 18-3

Quality of the Fit of the Absolute Radiant Flux Data to the Standard Nonlinear Response Equation

REFERENCE NO.	WAVELENGTH LAMBDA (MICRO M)	WAVENUMBER (1/CM)	STANDARD DEVIATION (PERCENT)
1	1.613	6200.012	11.001
2	1.600	6250.000	10.517
3	1.587	6300.006	10.653
4	1.575	6350.013	7.926
5	1.563	6400.000	7.384
6	1.550	6449.948	6.671
7	1.533	6499.837	5.913
8	1.527	6550.075	5.468
9	1.515	6599.789	4.833
10	1.504	6649.820	4.856
11	1.493	6700.167	4.822
12	1.481	6749.915	3.947
13	1.471	6799.946	4.126
14	1.460	6849.784	4.677
15	1.449	6899.883	3.613
16	1.439	6950.236	1.841
17	1.429	6999.860	3.060
18	1.419	7050.197	2.231
19	1.409	7099.751	2.787
20	1.399	7150.007	2.076
21	1.389	7199.942	3.750
22	1.379	7250.054	3.227
23	1.370	7299.803	7.515

TABLE 18-4

Normalized Inverse Spectral Areance (Irradiance) Responsivity

REFERENCE NO.	WAVELENGTH LAMBDA (MICRO M)	WAVENUMBER (1/CM)	RESPONSIVITY C LAMBDA (UNITLESS)
1	1.613	6200.012	6.810
2	1.600	6250.000	5.098
3	1.597	6300.006	4.063
4	1.575	6350.013	3.460
5	1.563	6400.000	2.972
6	1.550	6449.948	2.596
7	1.538	6499.837	2.287
8	1.527	6550.075	2.053
9	1.515	6599.789	1.843
10	1.504	6649.820	1.667
11	1.493	6700.167	1.508
12	1.481	6749.916	1.378
13	1.471	6799.946	1.269
14	1.460	6849.784	1.166
15	1.449	6899.383	1.129
16	1.439	6950.236	1.064
17	1.429	6999.860	1.000
18	1.418	7050.197	1.017
19	1.409	7099.751	1.096
20	1.399	7150.007	1.182
21	1.389	7199.342	1.231
22	1.379	7250.054	1.311
23	1.370	7299.803	1.389

TABLE 18-5

Spectral Areance (Irradiance) Calibration Equations (W/(SQCM-MICRO M))

CHANNEL	EQUATION
PREAMP	E LAMBDA = C LAMBDA * (2.036-13V + -2.180-19V**2)

18-3-1   Spectral Areance [Irradiance]

The absolute spectral areance [irradiance] of a point source is obtained using the calibration equation as

$$E(\lambda) = C(\lambda)[2.04 \times 10^{-13}V - 2.18 \times 10^{-19}V^2] \quad [\text{W cm}^{-2}\ \mu\text{m}^{-1}]$$

$$(18\text{-}1)$$

where V is the output in FFT units (see Table 18-5). The relative inverse spectral responsivity function $C(\lambda)$ is tabulated in Table 18-4 where it varies from unity upward. The value $C(\lambda) = 1$ corresponds to the wave number of maximum sensitivity for this system.

18-3-2   Dynamic Range

The standard deviation (rms value) of the dark-noise output was estimated at 16.8 FFT units. Substitution of this value into Eq. (18-1) yields the noise equivalent flux density NEFD for $C(\lambda) = 1$:

$$\text{NEFD} = 3.43 \times 10^{-12} \quad [\text{W cm}^{-2} \, \mu\text{m}^{-1}]. \tag{18-2}$$

Full-scale output corresponds to approximately 6464 FFT units which yields

$$E_{\text{max}} = 1.32 \times 10^{-9} \quad [\text{W cm}^{-2} \, \mu\text{m}^{-1}]. \tag{18-3}$$

18-3-3   Calibration Uncertainty

The calibration uncertainty is very difficult to estimate for an interferometer because the responsivity depends on modulation efficiency which in turn is dependent on mirror alignment. However, the data yielded

$$\sigma_{\text{L}} = \pm 4.7\% \qquad \text{(linearity: Table 18-2),} \tag{18-4}$$

$$\sigma_{\text{W}} = \pm 1.8\%\text{–}11\% \qquad \text{(absolute: Table 18-3),} \tag{18-5}$$

$$\sigma_{\text{N}} = \pm 16.8 \quad \text{FFT units} \quad \text{(system noise).} \tag{18-6}$$

The precision (repeatability) for measurements associated with this data run, for large signal-to-noise ratio (where σ_{N} can be neglected) varies from ± 5.03 to $\pm 12.0\%$ depending on the wavelength (see Table 18-4).

The accuracy depends not only on the standard source uncertainty, but also upon the polarization, reflectance, transmittance, and geometrical factors of collimator mirrors, windows, and apertures, all of which are difficult to qualify. Therefore no attempt is made to estimate the accuracy of this spectrometer.

REFERENCES

1 F. Gram and F. Costa, Determination of polarization in optical instruments and its metro-logical implications. *Appl. Opt.* **13**, 2228–2232 (1974).

2 K. D. Mielenz and K. L. Eckerle, Design, Construction, and Testing of a New High Accuracy Spectrophotometer. *Natl. Bur. Stand.* (*U.S.*), *Tech. Note* No. 729 (1972).

3 D. J. Baker, A. Steed, R. Huppi, and K. Baker, Twilight transition spectra of atmospheric O_2 ir emissions. *Geophys. Res. Lett.* **2**, 235–238 (1975).

4 D. J. Baker, W. Pendleton, Jr., A. Steed, and R. Huppi, Near-infrared spectrum of an aurora. *J. Geophy. Res.* **82**, 1601–1609 (1977).

APPENDIX

A

SI Base Units

Entity	Term	Symbol
Length	meter	m
Mass	kilogram	kg
Time	second	sec
Electric current	ampere	A
Thermodynamic temperature	kelvin	K
Amount of substance	mole	mole
Luminous intensity	candela	cd

Of the SI base units, the one of particular interest in the branch of photometry is the candela, defined as follows: "The candela is the luminous intensity, in the perpendicular direction, of a surface of 1/600,000 square meter of a blackbody at the temperature of freezing platinum under a pressure of 101,325 newtons per square meter.*"

* E. A. Mechtly, "The International System of Units," Rep. NASA P-7012, pp. 4, 5. Off. Tech. Util., NASA, Washington, D.C. (1969).

APPENDIX

B

SI Prefixes

Factor	Prefix	Symbol	Factor	Prefix	Symbol
10^{12}	tera	T	10^{-2}	centi	c
10^9	giga	G	10^{-3}	milli	m
10^6	mega	M	10^{-6}	micro	μ
10^3	kilo	k	10^{-9}	nano	n
10^2	hecto	h	10^{-12}	pico	p
10^1	deka	da	10^{-15}	femto	f
10^{-1}	deci	d	10^{-18}	atto	a

The terms and symbols listed in the tabulation are used, in combination with the terms and symbols, respectively, of the SI units, as prefixes to form decimal multiples and submultiples of those units.*

* E. A. Mechtly, "The International System of Units," Rep. NASA AP-7012, p. 3. Off. Tech. Util., NASA, Washington, D.C. 1969.

C

Atomic Constants*

Entity	Symbol	Value	Error (ppm)	Prefix	Unit
Speed of light in vacuum	c	2. 997 925 0	0.33	$\times 10^8$	m sec^{-1}
Gravitational constant	G	6. 673 2	460	10^{-11}	N m^2 kg^{-2}
Avogadro constant	N_A	6. 022 169	6.6	10^{26}	kmole^{-1}
Boltzmann constant	k	1. 380 622	43	10^{-23}	J K^{-1}
Gas constant	R	8. 314 34	42	10^3	J kmole^{-1} K^{-1}
Volume of ideal gas (standard conditions)	V_0	2. 241 36		10^1	m^3 kmole^{-1}
Faraday constant	F	9. 648 670	5.5	10^7	C kmole^{-1}
Unified atomic mass unit	u	1. 660 531	6.6	10^{-27}	kg
Planck constant	h	6. 626 196	7.6	10^{-34}	J sec
	$h/2\pi$	1. 054 591 9	7.6	10^{-34}	J sec
Electron charge	e	1. 602 191 7	4.4	10^{-19}	C
Electron rest mass	m_e	9. 109 558	6.0	10^{-31}	kg
		5. 485 930	6.2	10^{-4}	u
Proton rest mass	m_p	1. 672 614	6.6	10^{-27}	kg
		1. 007 276 61	0.08	—	u
Neutron rest mass	m_n	1. 674 920	6.6	10^{-27}	kg
		1. 008 665 20	0.10	—	u
Electron charge-to-mass ratio	e/m_e	1. 758 802 8	3.1	10^{11}	C kg^{-1}
Stefan–Boltzmann constant	σ	5. 669 61	170	10^{-8}	W m^{-2} K^{-4}
First radiation constant	$8\pi hc$	4. 992 579	7.6	10^{-24}	J m
Second radiation constant	hc/k	1. 438 833	43	10^{-2}	m K
Rydberg constant	R_∞	1. 097 373 12	0.10	10^7	m^{-1}
Fine structure constant	α	7. 297 351	1.5	10^{-3}	
	α^{-1}	1. 370 360 2	1.5	10^{+2}	
Bohr radius	a_0	5. 291 771 5	1.5	10^{-11}	m
Magnetic flux quantum	Φ_0	2. 067 853 8	3.3	10^{-15}	Wb
Quantum of circulation	$h/2m_e$	3. 636 947	3.1	10^{-4}	J sec kg^{-1}
	h/m_e	7. 273 894	3.1	10^{-4}	J sec kg^{-1}

* E. A. Mechtly, "The International System of Units," Rep. NASA AP-7012, pp. 7, 8. Off. Tech. Util., NASA, Washington, D.C. (1969).

Index

A
B
C 8
D 9
E 0
F 1
G 2
H 3
I   4
J 5